Backyard Chickens for Beginners in the Suburbs

A First-Time Chicken Owner's Guide to Legal Small Flocks, Safe Coops and Runs, Odor Control, Predator Protection, and Fresh Eggs at Home

Jennifer H. Johnson

Backyard Chickens for Beginners in the Suburbs is written for educational and informational purposes only. While every effort has been made to present accurate, practical, and carefully researched material, no warranties of any kind are declared or implied. The author and publisher assume no responsibility for errors, omissions, changes in laws or regulations, differences in local enforcement, product failures, animal behavior, weather events, predator activity, illness, injury, property damage, nuisance complaints, or any other consequence resulting from the use or application of the information contained in this book.

By reading this book, the reader acknowledges that the author and publisher are not responsible for any direct or indirect loss, damage, injury, expense, legal dispute, flock loss, animal illness, neighbor conflict, or other outcome related to backyard chicken keeping, coop construction, animal care, local rule compliance, or the reader's personal decisions.

Legal and Local Ordinance Notice

This book is a general beginner's guide to planning and keeping a small flock of backyard hens in a suburban setting. It is not legal advice and should not be treated as a substitute for consulting qualified legal counsel, local officials, zoning departments, animal-control offices, homeowner association boards, landlords, or other authorities with jurisdiction over a specific property.

Backyard chicken rules vary widely by city, county, township, zoning district, homeowner association, deed restriction, lease agreement, neighborhood covenant, and local nuisance ordinance. Rules may also change after publication. It is the reader's responsibility to verify current laws, permits, flock limits, rooster restrictions, coop setbacks, waste rules, egg sale rules, building codes, and neighborhood requirements for the reader's exact address before purchasing birds, supplies, materials, or a coop.

Nothing in this book grants permission to keep chickens, construct a coop, sell eggs, compost manure, modify property, install fencing, or use any product or method where doing so is prohibited, restricted, or regulated by applicable law, contract, covenant, or local authority.

Animal Health and Veterinary Notice

This book is not veterinary advice and is not a substitute for professional diagnosis, treatment, medication guidance, or emergency animal care. Backyard poultry can experience illness, injury, parasites, nutritional problems, heat stress, frostbite, predator trauma, reproductive issues, and contagious disease. Readers should consult a qualified poultry-aware veterinarian, university Extension service, state veterinarian, diagnostic laboratory, or other qualified animal-health professional when birds appear sick, injured, distressed, or abnormal.

This book does not prescribe medications, dosages, home remedies, vaccines, treatments, withdrawal times, or disease-control protocols. Any medication, supplement, pesticide, disinfectant, feed additive, or treatment product should be used only according to current label directions, veterinary guidance, and applicable laws.

Public Health and Biosecurity Notice

Backyard poultry may carry organisms that can cause illness in humans, including children, older adults, pregnant women, and people with weakened immune systems. Readers are responsible for following current public-health guidance from appropriate authorities, including but not limited to guidance related to Salmonella, Campylobacter, avian influenza, manure handling, handwashing, poultry contact, egg handling, cleaning practices, and biosecurity.

Children should be supervised around poultry, coops, feed, bedding, manure, and eggs. Poultry should not be brought into kitchens, dining areas, bedrooms, bathrooms, or other human living and food-preparation spaces. Readers should verify and follow current recommendations from the CDC, USDA APHIS, state agriculture departments, local health departments, university Extension offices, and qualified poultry-health professionals.

Construction, Utility, and Property Safety Notice

This book discusses chicken coops, runs, fencing, hardware cloth, latches, predator aprons, feed storage, drainage, weather protection, and other backyard infrastructure for general planning purposes only. It is not a construction manual, engineering guide, electrical guide, building-code reference, or substitute for professional design, inspection, or installation.

Readers are responsible for ensuring that all structures, fencing, electrical components, lighting, heaters, water systems, trenches, posts, foundations, roofs, and other installations comply with manufacturer instructions, local building codes, utility requirements, safety standards, and applicable permits.

Before digging, trenching, driving posts, installing predator aprons, modifying soil, or disturbing the ground in any way, readers should contact 811 or the appropriate local utility-locate service and follow all required procedures. Utility locating may not identify private lines, irrigation systems, invisible dog fences, septic components, drainage systems, or other buried hazards. Readers are responsible for taking appropriate precautions and hiring qualified professionals when needed.

Predator, Pet, and Containment Notice

No coop, run, fence, latch, automatic door, hardware cloth, predator apron, or management system can be guaranteed completely predator-proof, dog-proof, escape-proof, weather-proof, rodent-proof, or failure-proof. Wildlife, domestic dogs, cats, rodents, snakes, weather events, falling limbs, human error, and product defects can all create risks.

For permissions, bulk sales, or special inquiries, contact: Support@ProfessionalSkillsPublishing.com

Contents

Introduction: Read this before you bring chickens home

PICTURE YOUR BACKYARD. YOU can probably see the vision in your mind: a sturdy little coop tucked into the corner and three or four hens scratching softly in the grass. But look a little closer at your actual yard. There is the property line, just ten feet from where you want the coop, bordering your neighbor's patio. Over the other fence is a high-energy dog. Tucked in a drawer sits an unread homeowner's association rulebook. The narrow side yard is too shady, and that perfect corner of the lawn turns deeply muddy every time it rains.

Keeping hens in the suburbs is less like planting a new garden and more like managing a small, daily household system in a shared neighborhood. This book exists to help you build that system. Our central goal is to help you plan a flock that meets exactly five criteria: **legal, safe, clean, contained, and neighbor-friendly.**

Many chicken resources are broad, focusing on homesteading philosophy, breed history, or large-scale flock management with the assumption that you have wide-open space. We are taking a different approach. This guide focuses entirely on suburban and neighborhood readiness. It is designed for cautious beginners, families, dog owners, and avid gardeners.

Above all, it is for readers who have not actually purchased birds yet. Before a single hen comes home, you need to understand exactly what your property can support. You need to verify local rules, map your yard, establish predator resistance, and plan for secure feed storage, daily manure management, and your immediate neighbor's peace of mind. Caution is not a failure here. If you read this book and decide the answer right now is "not yet," that is a highly successful and responsible outcome.

Before we go further, we need to clear away a few common misconceptions. First is the idea that chickens are naturally easy, so you can just figure it out later. Trying to navigate local zoning rules, fix drainage issues, or manage a predator breach after you already have live animals in your yard is incredibly stressful. Second is the assumption that because a friend a few towns over keeps chickens without a problem, you probably can too. Rules vary wildly by municipality, county, zoning district, deed restriction, and HOA.

Third is the belief that if you build an incredibly cute coop, your neighbors will automatically tolerate unexpected noise, flies, or smells. Aesthetics do not cancel out nuisance complaints. And finally, do not assume backyard eggs will instantly save you money at the grocery store. Keeping a healthy suburban flock is a rewarding pursuit, but it requires upfront financial investment, an emergency fund, and ongoing daily upkeep. There is no such thing as a cheap backyard egg.

To keep our focus sharp, we need to establish exactly what this book covers. This is a U.S.-focused suburban readiness planner for small flocks of laying hens. We will use the regulatory language common to the United States, referencing guidance from the CDC, USDA APHIS, university Extension offices, state veterinarians, and 811 utility-locate services.

Because we are designing a neighborhood-friendly system, several topics are entirely out of scope. This book does not cover meat birds, slaughter, breeding, hatching, or fertile eggs. It also strongly discourages keeping roosters, as they are rarely appropriate, or legal, for beginner suburban environments.

Furthermore, this book is a readiness planner, not a professional manual. It does not provide formal legal counsel, veterinary diagnoses, electrical wiring instructions, or detailed construction engineering plans. You must always verify current local rules with the relevant local authority. Outdoor electrical setups require strict adherence to manufacturer instructions and local safety codes. Likewise, sick or injured birds require advice from a qualified poultry-aware veterinarian, Extension service, or diagnostic lab. This book will not tell you how to dose medications or prescribe home remedies. We provide the framework; you must provide the local verification.

This book will walk you through a step-by-step process: verifying local rules, mapping your yard, choosing a starter flock pathway, whether chicks, pullets, or adult hens, and organizing your daily routines. You will finish Part 1 with a fully drafted first-flock operating plan for your household.

To begin, take a moment right now, before you spend a single dollar on supplies, lumber, or birds, to write down your three biggest fears about keeping chickens. Whether you are worried about attracting rodents, dealing with neighbor complaints, or managing winter weather, name them. We will address them practically as you build your plan.

Welcome to Chickens... in the Suburbs!

Jenny Johnson (Suburban Chicken Momma)

Suburban Chickens

Part 1: Decide before you buy birds

SPRINGTIME AT THE LOCAL farm store is a dangerous place for a suburbanite with a backyard and a dream. You walk in looking for grass seed or work gloves, and then you hear it: the soft, unmistakable peeping echoing from the back aisle. You wander over. Under the warm glow of heat lamps sit galvanized troughs filled with dozens of tiny, day-old chicks. They tumble over each other and scratch at fresh pine shavings. A handwritten sign taped to the edge of the trough says something simple: "Assorted Pullets: $4.99 each."

It feels deceptively easy. For a twenty-dollar bill, you could walk out of the store with the living foundation of your future flock. The cashier is happy to sell you starter feed, a plastic waterer, and a heat bulb. You imagine keeping the birds in a cardboard box in the garage for a few weeks while you order a beautiful coop online.

If you have been tempted by this exact scenario, you are in excellent company. Thousands of backyard flocks start with a spontaneous springtime trip to the store. The homesteading fantasy is powerful, especially when it costs less than five dollars a bird. In a residential neighborhood setting, however, an impulse poultry purchase usually leads to a stressful, legally compromising situation.

Those tiny chicks grow significantly faster than most beginners expect. Within a few short weeks, fluffy down gives way to awkward, dusty feathers. Soon they are testing their wings, kicking pine shavings out of their makeshift brooder, and generating a surprising volume of waste and odor. They will quickly outgrow the plastic storage tote in your guest bathroom or the cardboard box in your mudroom long before that mail-order coop arrives on your front porch or the ground thaws enough to build a permanent run.

Also, bins labeled "assorted pullets" are never fully guaranteed. Depending on hatchery sexing methods and human error, there is always a risk of accidentally bringing home a rooster, a reality that suburban zoning rarely forgives.

More importantly, bringing birds home early forces you into a permanent reactive scramble. Rushing to accommodate rapidly growing birds leaves no time to carefully research zoning laws or check homeowner's association guidelines to see if poultry is explicitly banned. You cannot thoughtfully survey your neighbors to gauge their comfort level or address concerns about noise and pests. Instead, you simply try to keep the birds alive, clean, and contained while hoping nobody complains.

In a rural setting, a space miscalculation or a delayed coop build means you simply patch together a larger temporary run out back over the weekend. The stakes are low. In the suburbs, the stakes are entirely different. A miscalculation often results in a formal warning from code enforcement, a compounding daily fine, a fractured relationship with the family next door, or a sudden order to rehome the birds.

Some keepers call these hidden flocks "outlaw chickens." Kept in secret behind tall privacy fences, these birds belong to owners violating municipal codes, health department regulations, or restrictive HOA covenants. While a few people might brag on internet forums about getting away with it, living with outlaw chickens is an exhausting, anxiety-filled experience.

With illegal birds, you cannot truly relax. You panic every time a hen sings her surprisingly loud daily egg song. You worry when a city utility truck parks on your street or when you invite guests over for a backyard barbecue, fearing someone might accidentally mention the coop.

That is why we are hitting the pause button right now. This section of the book exists to make sure you do not become an outlaw chicken keeper by accident. Before you spend a single dollar on feed, before you pick out a coop design, and absolutely before you bring a live animal onto your residential property, you must ensure your environment can legally and practically support them.

Suburban chicken keeping is a long game. A flock of healthy laying hens can easily live for five to eight years, often well past their peak laying years. Doing everything above board is the only way to guarantee long-term peace of mind. When you follow the rules and plan ahead, you can sit in your yard with your morning coffee, watch the hens scratch in the dirt, and know with absolute certainty that they are allowed to be there.

Part 1 of this book is not an instruction manual for raising birds. It is a decision matrix. Your first job is not breed selection, coop design, or predator proofing. Your first job is deciding whether chickens actually fit your household, your local rules, your budget, your neighbors, and your specific property layout.

As you work through the upcoming chapters, you are looking to arrive at one of three distinct conclusions: Yes, No, or Not Yet.

A "Yes" means you have the legal right, the financial resources, the neighborhood capital, and the personal time to manage a daily household system safely and responsibly.

A "No" means you have uncovered a strict legal ban, a prohibitive budget reality, or a lifestyle conflict that makes keeping chickens unmanageable. Finding out it is a "No" before you spend money is a massive success.

A "Not Yet" means you want to keep chickens, but you need to wait for a different season of life. Perhaps you need to save more money for a high-quality predator-resistant run, or maybe you need to wait until you move out of an HOA-governed subdivision. A "Not Yet" protects you from the stress of a failed flock.

To reach your final decision, the chapters ahead will guide you through four specific gates of suburban readiness.

First, we assess your household readiness. We look at what a backyard flock actually demands of your daily schedule, physical space, and routine, helping you decide if you have the margin to care for livestock every single day, regardless of the weather.

Second, we walk through the process of securing legal permission. You will learn exactly how to build your "legal stack," researching your specific municipality, county, zoning ordinances, and deed restrictions. We will not teach you the laws themselves; we will teach you how to find the specific rules governing your exact address.

Third, we address the reality of living close to other people. You will prepare for the neighborhood impact, learning how to evaluate how a coop affects the people who share your property lines and how to handle inevitable concerns about noise and smell.

Finally, we strip away the homesteading myths and build a realistic budget. We will establish exactly why suburban eggs are never cheap, examining the actual costs of secure infrastructure, high-quality feed, and ongoing maintenance.

Our ultimate goal is to help you plan a flock that meets five uncompromising criteria: legal, safe, clean, contained, and neighbor-friendly. You want your backyard birds to be a source of daily joy, not an ongoing burden or a community dispute. That transformation begins with the courage to plan thoroughly. Let's figure out if you are truly ready to bring the birds home.

Chapter 1: The yes, no, or not yet decision

THE MOST SUCCESSFUL BACKYARD chicken keepers are not the people who rush. They do not walk into a farm store on a whim, fall in love with a trough of fluffy chicks, and bring them home in a cardboard box to figure things out on the fly. Instead, the people who build manageable, enjoyable flocks tend to do one thing early: they notice problems while those problems are still cheap and simple to solve.

Excitement is an excellent starting point, but it is a terrible operating plan. Excitement will get a beautiful coop built on a sunny Saturday, but it will not trudge out in the freezing rain to break ice out of a waterer in January. Excitement will not convince a homeowner's association to drop a fine, nor will it magically train a high-prey-drive dog to ignore a fluttering hen.

It is tempting to view a backyard flock as a fun gardening project or an interactive yard decoration. But chickens are a daily livestock commitment, scaled down to fit your neighborhood. Bringing them into a suburban environment requires shifting how your residential property functions. Your first decision is not picking a breed or choosing a coop color. Your first decision is figuring out whether your household can safely and practically support a flock at all.

This chapter gives you permission to pause, evaluate, and even walk away without feeling like you failed. "Not yet" is a highly responsible answer, and knowing when to wait is a vital part of keeping a healthy flock.

The responsible outcomes

As you consider bringing chickens onto your property, your goal is to reach one of three conclusions: a verified yes, a definitive no, or a strategic not yet.

A verified yes means your commitment is documented, not merely enthusiastic. You have checked the local laws in writing, you have mapped your yard, you have budgeted for secure materials, and you have achieved a consensus within your household. A "yes" means you are ready to execute a clear plan.

A definitive no means you have found an unresolvable mismatch between your current living situation and the realities of a flock. Saying no is not a defeat; it is a profound sign of respect for the animals and for your own peace of mind.

A strategic not yet means you want to keep chickens but have hit a solvable barrier. The legal rules might require a permit you have not applied for, your budget might need time to grow, or your yard might need regrading to fix a drainage issue. "Not yet" is a practical, responsible position. You are prioritizing a safe, secure setup over impatience.

The major readiness categories

Before you purchase feed or look at coop designs, you must evaluate your household against eleven major readiness categories.

1. Law. Are hens allowed at your specific address? This requires checking municipal ordinances, county zoning laws, homeowner association covenants, and lease agreements. It also includes understanding property line setbacks that dictate exactly where a coop can legally sit.

2. Household agreement. Are all the adults in the home on board? A flock cannot be a secret, and the daily chores should not be forced on a resentful partner or roommate.

3. Neighbors. Will your setup unnecessarily burden the people living next door? Successful suburban chicken keeping requires placing the coop where it minimizes noise, odor, and visual impact on your neighbors' outdoor living spaces.

4. Money. Can the setup and upkeep be paid for without financial strain? Chicks may be cheap, but secure coops, heavy-duty hardware cloth, and ongoing feed costs add up quickly.

5. Yard. Do you have a practical location with good drainage so the birds are not standing in mud, adequate shade for hot summers, and easy human access for hauling heavy feed bags?

6. Pets. If you have dogs or outdoor cats, how will you manage them? A high-prey-drive dog requires permanent, fail-proof physical separation from the flock.

7. Children. Can you establish and enforce boundaries? Children require adult supervision around livestock to prevent injury to the birds and to manage public health risks like Salmonella.

8. Time. Who owns the daily routine? Chickens must be let out, locked up, fed, and watered daily. You also need a plan for dependable vacation care when you travel.

9. Predators. Are you prepared to build a heavily fortified run? Every neighborhood has predators, from hawks and raccoons to roaming domestic dogs. Strong containment is not optional.

10. Hygiene. Are you willing to manage animal waste, enforce handwashing, and adopt a strict policy of keeping poultry gear and birds out of human food preparation areas?

11. Tolerance for loss or inconvenience. Can you handle a muddy yard, a ruined flowerbed, or the sudden loss of a bird? Livestock ownership inevitably involves mess, illness, and occasional heartbreak.

The walk-away or wait triggers

Some barriers are absolute. If you encounter one of these, you should immediately trigger a "no" or a "not yet." Do not bring chickens onto your property if any of the following statements apply:

- **The rules prohibit hens.** Local laws, your HOA, or your lease explicitly forbid keeping backyard poultry at your address.

- **Your plan depends on hiding the birds.** Relying on secrecy from landlords or neighbors is stressful, eventually fails, and usually results in the rushed, unfair rehoming of the flock.

- **No adult owns the daily care.** Children can be wonderful helpers, but an adult must hold ultimate responsibility for the health, safety, and hygiene of the animals.

- **Dog risk is unmanaged.** If you have a dog that chases wildlife and you have no realistic way to build a completely separate, heavily fortified enclosure, do not get chickens.

- **You refuse to lock them up.** Letting birds wander an unfenced suburban yard without a predator-resistant run or a secure nighttime coop simply sets up a buffet for local wildlife.
- **Your budget is strained.** If paying for a secure lock, buying a bag of feed, or replacing a damaged waterer will cause financial hardship, wait.

The yes, no, or not yet decision matrix

Now it is time to capture a snapshot of your current readiness. You do not need to overthink this, but you do need to be honest.

For each category, ask yourself the following questions and assign yourself a Yes, a No, or a Not Yet.

- **Law:** Do I have written verification that I am legally allowed to keep chickens at this exact address?
- **Household:** Do all adults in the home agree to this project and understand the daily commitment?
- **Neighbors:** Have I considered the impact of a coop on my immediate neighbors?
- **Money:** Do I have the funds for a highly secure setup and ongoing feed without financial strain?
- **Yard:** Do I have a dry, accessible space for a coop that complies with local setbacks?
- **Pets:** Do I have a solid, physical plan to manage my dogs around the flock?
- **Children:** Am I ready to enforce strict handwashing and supervision rules for kids?
- **Time:** Can I commit to a morning and evening routine, every single day?
- **Predators:** Am I willing to build or buy a highly secure enclosure with hardware cloth?
- **Hygiene:** Am I prepared to actively manage manure and keep poultry supplies out of my kitchen?
- **Tolerance:** Am I mentally prepared for yard damage, mess, and potential flock loss?

If your answers are entirely "Yes," you are cleared to move forward. If you have a "No" in a critical, unchangeable area like Law or Household Agreement, your final answer is "No." If you have a mix of "Yes" and "Not Yet," you have a project list.

A completed example: Choosing "not yet" responsibly

To see how this tool works in practice, consider a cautious household, the Miller family, that used this matrix and ultimately decided on "not yet."

The Millers had plenty of space and a strong desire to start a flock. When they sat down with the decision matrix, they quickly checked off several boxes. They called their township and confirmed that they were legally allowed to keep up to six hens, assigning a Yes to the Law category. They had enough savings to cover a high-quality coop, securing a Yes for Money. Both adults were fully on board, securing a Yes for Household, and they knew exactly where the coop would go in the yard.

But when they reached the Pets category, they hesitated. They had adopted a young terrier mix six months prior. The dog had a high prey drive and spent his afternoons frantically chasing squirrels along the fence line. The Millers realized that their current chain-link fence was too low, and they had not yet figured out how to physically separate the dog's play area from the future chicken run.

When they looked at the Time category, they found another hurdle. They were scheduled to travel extensively for youth sports tournaments throughout the upcoming summer, and they had not yet found a reliable pet sitter who was comfortable handling poultry.

Instead of ignoring these issues and hoping for the best, the Millers marked "Not Yet" for Pets and Time. Their overall decision became a strategic "Not Yet."

This was a massive success. The emotional permission to wait allowed them to replace their fencing that autumn and arrange for a neighbor to help with pet sitting the following year. By delaying their purchase by twelve months, they saved themselves the stress of a dog-related tragedy and the panic of finding last-minute vacation care. When they finally brought home their pullets the next spring, they did so with absolute confidence.

Your next step

To clarify your own position, take a quiet moment and answer these two final questions:

1. What would have to be true for my situation to become a responsible "yes"?
2. What is the one specific thing I need to verify before I spend any money?

If you have decided that a "not yet" or a "no" is the best answer for your household right now, take a breath. You have saved yourself a tremendous amount of stress, money, and heartache. You have made the right choice, and you can close this book knowing you did the responsible thing.

If you are looking at a "yes," or if your "not yet" simply requires a little more research to clear a solvable barrier, you are ready to move forward. The excitement is still there, but now it is backed by strategy. Your very first operational step is to remove the guesswork from the legal gate. In the next chapter, we are going to lock down the exact rules for your property so you can build your flock on an unshakable foundation.

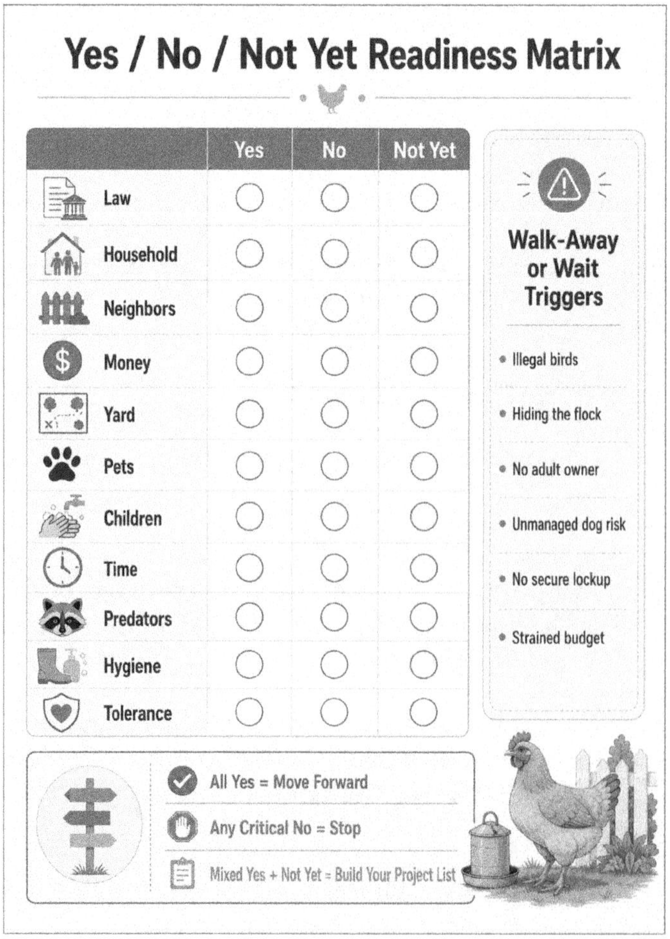

Chapter 2: Verify the rules before you buy birds

TWO FRIENDS, DAVID AND Abigail, live just three miles apart in the same county. Abigail has kept a flock of six hens for three years. She regularly brings David cartons of fresh eggs and talks about how her garden has improved thanks to the compost.

Inspired, David decides it is time to start his own flock. He visits the local farm store, buys a brooder full of day-old chicks, and spends his weekend building a coop against his back fence. A week later, an official letter arrives in David's mailbox. It is a notice of violation from his homeowner's association, accompanied by a warning from city code enforcement. He has fourteen days to remove the chickens and tear down the unapproved coop.

David is stunned. How can this happen when Abigail lives just down the road and has kept chickens for years without a single complaint?

The answer is the single most important lesson of suburban chicken keeping: chicken rules are local. Abigail lives in an unincorporated part of the county with no homeowner's association. David lives inside city limits within a deed-restricted neighborhood. They share a zip code, but their legal realities are entirely different.

When you are excited to start a new hobby, it is easy to rely on assumptions. You might hear someone say, "My friend has chickens, so they must be legal here." You might assume that because a local farm store sells chicks every spring, anyone in the area can buy them. But farm stores operate under commercial retail rules and sell to a wide region. They do not know the zoning ordinances of your specific address, nor do they know the private contracts you signed when you bought or rented your home.

This chapter gives you a structured process for verifying the rules before you spend money on birds or equipment. You are going to document exactly what is allowed on your property so you have a clearer record if questions ever come up.

* *A note on legal information:* ***This chapter provides a framework for organizing your research, communicating with local officials, and documenting the answers you receive. It does not constitute legal advice. There is no universal backyard chicken law, and local ordinances change frequently. The author and publisher cannot guarantee that following this process will secure legal permission for your flock. It is entirely your responsibility to verify current local laws, secure necessary permits, and comply with all public ordinances and private contracts associated with your specific property. If you encounter complex lease agreements, deed restrictions, or municipal codes, consult a qualified attorney in your jurisdiction.*** *

The legal stack

There is no single "chicken law" that governs your backyard. What is legal for you depends on a highly specific stack of local authorities. Depending on your address, your property might be governed by city ordinances, county health codes, zoning districts, or animal-control regulations. You could also be subject to nuisance laws, building codes, homeowner's association covenants, or a landlord's lease agreement.

Reference **Appendix A: Local-Law Verification Record** as you go through this chapter.

Think of these authorities as a stack of filters. A "yes" from one layer does not automatically cancel out a "no" from another. For example, a county ordinance allowing chickens does not override a private HOA covenant banning them. Because of this overlapping authority, you must clear your plans with every level of governance that applies to your property.

Here are the primary layers of the legal stack you must navigate:

Municipal, town, or county codes: If you live inside city or town limits, municipal codes dictate what you can do on your property. If you live outside city limits, county or township codes apply. These codes broadly establish whether keeping backyard poultry is a permitted use of residential land.

Zoning and planning departments: Zoning boards dictate the physical structures on your property. They categorize a chicken coop as an "accessory structure," similar to a tool shed or a detached garage. Zoning rules determine how large that structure can be, how tall it can be, and exactly where it can sit in your yard.

Animal control: While frequently associated with managing stray dogs and cats, local animal-control offices often enforce municipal livestock rules. They are typically the agency that monitors flock-size limits, animal-welfare standards, and noise complaints.

Private contracts (HOAs, leases, and deed restrictions): This is the most restrictive layer of the stack. Even if a city actively encourages backyard flocks, a private contract can ban them entirely. If you live in a managed community, your Homeowner's Association (HOA) covenants legally bind you. If you rent, your lease dictates your property use. Even if you own an older home without an active HOA, your property may be bound by historical deed restrictions that prohibit livestock. If a private contract appears to restrict poultry, treat that as a stop sign until you have written clarification.

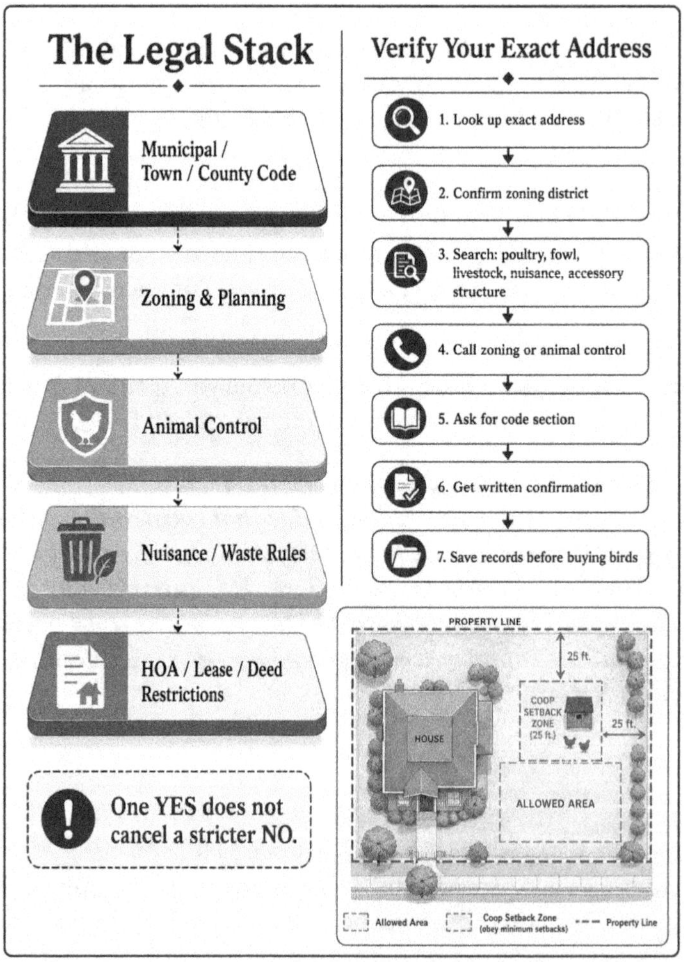

The rule categories to verify

Do not assume that because you only want three quiet hens, the rules do not apply to you. Before you move forward, verify the facts across several distinct categories.

Flock limits and rooster rules: Find out exactly how many birds you are allowed to keep. Municipalities rarely grant unlimited permission for residential lots. They typically cap the flock at a specific number, such as four or six birds, or scale the allowance based on your exact lot square footage. Additionally, verify the rules regarding roosters. Because of their noise, many suburban areas explicitly ban roosters while permitting hens.

Coop placement and setbacks: Setbacks are the legally required distances between a structure and a property line, a fence, a road, or a neighboring residence. Zoning codes will dictate exactly how far your coop must be from these boundaries. Do not fall into the trap of thinking you can just move the coop later if someone complains. Quality coops are incredibly heavy, and moving one safely is difficult. Know exactly where the coop is legally allowed to sit before you build or buy it.

Permits and inspections: Look for permit, inspection, or registration requirements. Some cities require you to pay a small fee and register your flock annually with animal control. Others require a formal building permit before you construct a coop. In some historic districts, any structure visible from the street requires a special review from a preservation board.

Waste, compost, and nuisance rules: Municipalities have a vested interest in public health. There may be specific local ordinances dictating how to manage soiled bedding and manure to prevent odor, flies, and rodents. You must also check for general nuisance laws. Nuisance codes are often broadly written, allowing authorities to issue citations for excessive noise, foul odors, or animals roaming off your property.

Egg sharing and sales: Your local zoning board may allow you to keep hens for personal egg consumption, but explicitly forbid you from running a commercial business from your home. If you plan to sell eggs, even just a few dozen a month to neighbors, you are stepping into a different regulatory category. State agriculture departments and local public health boards maintain specific food-safety rules governing egg distribution. Depending on your state and locality, these rules may address whether and how eggs must be washed, refrigerated, labeled, packaged, or otherwise handled for sale.

How to ask clear questions

Now that you understand the categories, you need a clear verification workflow. Your goal is to find the person with the authority to give you a definitive answer, and to get that answer backed up by documentation.

Before you make a phone call or type an email, look up your exact property address. This is the most critical piece of information you have. Zoning rules can change from one side of a street to the other. When you speak to a local official, do not ask, "Are chickens allowed in this town?" Ask, "Are chickens allowed at this specific address?"

Start from the top down. Many cities and counties publish their municipal codes online in searchable databases. Search for terms like "poultry," "fowl," "livestock," "accessory structures," and "nuisance." Read the code yourself to establish a baseline understanding before you reach out to agencies.

When you contact your local zoning office or animal-control department, listen carefully for specific numbers and code sections. Ask them to confirm your zoning district, the allowable flock limits, the exact coop setback measurements, and any permit requirements. Always ask them to point you to the specific section of the published ordinance so you can read it yourself.

If you are communicating with an HOA management company or a landlord, your inquiry must be equally precise. State your address and ask clearly if backyard poultry is permitted under current covenants or lease terms. If it is permitted, ask what specific architectural review process is required for placing a small coop in the backyard.

During your research, you may come across resources from your local university Extension office or state animal-health department. Extension offices provide expert guidance on poultry care, but they do not grant legal permission to keep birds.

State animal-health resources will alert you to disease outbreaks, but they do not override your local zoning laws. Keep your legal questions focused on the specific local agencies that hold jurisdiction over your address.

How to document your findings

Do not rely on your memory. You are going to gather a lot of overlapping information, and verbal conversations are easily forgotten or denied. A written record gives you something concrete to refer back to if questions come up, or if the personnel at a local agency change.

To make this easy, use the Local-law verification record provided in the companion PDF [Appendix A: Local verification packet]. This packet also includes suggested email and phone scripts to help you navigate conversations with zoning offices, animal control, and HOA boards.

For every code section you check and every conversation you have, document the following information:

- **Contact name:** Who exactly did you speak with?
- **Agency and role:** What department do they work for, and what is their title?
- **Date and time:** When did the conversation take place?
- **Exact answer:** What are the specific limits, setbacks, or permit requirements?
- **Source link:** What is the specific section number of the municipal code, or where is the policy written?
- **Follow-up:** What are your next steps?

If you have a phone conversation with a zoning official or an HOA representative, send a brief, polite follow-up email outlining what you discussed. For example: "Thank you for speaking with me today. To confirm my understanding of our conversation, my address is zoned residential, I am allowed up to six hens with no roosters, and my coop must be at least ten feet from the rear property line." This creates a time-stamped, written trail of the information you were given.

Before recording any phone call or in-person conversation with officials, neighbors, or landlords, check the recording-consent law that applies in your jurisdiction and, when in doubt, get clear permission first. Recording laws vary by jurisdiction, and recording without required consent can carry legal risk. For this process, stick to written documentation and polite email summaries.

What to do when answers conflict

Sometimes, you will get conflicting information. Let's look at a case study of a prospective chicken keeper named Elena.

Elena called her local animal-control office. The officer told her she was allowed to keep up to six hens. But because she was following a thorough verification process, she also called her city planning department. The planner informed her that while hens were indeed legal under animal-control rules, her specific neighborhood was part of a historic overlay district. Any new structure visible from the street, including a chicken coop, required a special review from the historic preservation board.

If Elena had stopped after calling animal control, she would have built a coop and likely received a citation for building an unapproved structure in a historic district.

When answers conflict or overlap, never assume that the easiest answer controls your property. For planning purposes, assume the most restrictive applicable requirement controls until the responsible office confirms otherwise.

If a local official gives you an answer over the phone that contradicts the written code you found online, politely ask for clarification. You can say, "Thank you for that information. I am looking at Section 8 of the municipal code online, which seems to say something slightly different regarding property setbacks. Can you help me understand how this code applies to my specific address?"

Who cannot give you legal permission?

As you verify your local rules, avoid the temptation to rely on shortcuts. The following sources cannot grant you legal permission to keep chickens on your property:

Social media groups: A neighborhood forum is a great place to find a local plumber, but it is a terrible place to get legal advice. The person commenting that chickens are perfectly fine in your town does not know your exact property lines, your specific zoning district, or the terms of your lease.

Feed store employees: Farm supply stores sell livestock to a wide geographical region. The cashier selling you a bag of chick starter does not know the covenants of your specific homeowner's association.

Coop builders and sellers: A local carpenter building coops and selling them online is running a construction business, not a zoning advisory service. It is entirely your responsibility to know if that product meets your local height and setback requirements.

Neighbors: Just because the person across the street has chickens does not mean they are legal. Your neighbor may be operating illegally; they may be ignoring HOA warnings, or they may have a special agricultural exemption that does not apply to your lot.

The "Do Not Buy Yet" triggers

As you work through your research, keep a strict boundary in place. If you encounter any of the following triggers, hit pause. Do not buy chicks, pullets, hens, or a coop until the issue is entirely resolved.

- **You have unresolved conflicting answers.** If animal control says yes but zoning says no, do not proceed until you have written clarification on which rule governs your property.
- **Your HOA approval is pending.** If your HOA requires architectural review for a coop, do not buy materials or birds until the board has formally signed off on your submitted plans.
- **Your lease is vague.** If your lease bans "livestock" but does not explicitly mention poultry, do not assume you have found a loophole. Get written approval from your landlord specifically allowing backyard hens.
- **You only have verbal approvals.** A verbal "go ahead" from an official or landlord is insufficient. If they will not put the permission in an email or a formal document, consider the matter unresolved.
- **You cannot meet the setback requirements.** If the zoning code requires a twenty-foot setback from property lines, but your backyard is only thirty feet wide, zoning rules effectively prevent you from keeping a flock.

Legal permission is only the first half of suburban chicken keeping. The law tells you what you are allowed to do, but your neighborhood relationships dictate how enjoyable the experience will actually be. Once your local law verification record is complete and all answers are confirmed, you are ready to look over the fence and create a neighbor-friendly plan.

Chapter 3: Make a neighbor-friendly promise you can keep

Most people do not have a problem with chickens in theory. When a neighbor hears you are thinking of starting a backyard flock, they might picture storybook farmyards, quaint wooden coops, and fresh eggs. But neighbors do not live next to theories. They live next to your actual backyard. What neighbors genuinely have a problem with is a sudden invasion of flies on their summer patio, a foul ammonia smell drifting over the fence, or a tarp-covered shack they have to look at every time they wash their dishes.

As covered in the previous chapter, verifying your local laws is non-negotiable. But a crucial lesson of suburban chicken keeping is that **legal does not automatically mean neighbor-friendly**. Local ordinances might technically allow you to build a chicken run three feet from your property line, but if that three feet is directly upwind of your neighbor's outdoor dining table, legal placement will still result in miserable neighbors and predictable conflict. Legal compliance keeps code enforcement off your back; neighbor-friendly management keeps peace over the fence.

When you bring livestock into a residential area, it is easy to adopt a defensive mindset. You might think, "It is my property, so my neighbors' opinions do not matter." But in a suburban setting, the air you share, the fences you share, and the sightlines you share mean your property does not exist in a vacuum.

A truly neighbor-friendly flock is defined by seven traits. It is:

1. **Clean:** Managed with strict sanitation protocols to prevent pest attraction.
2. **Contained:** Kept within secure boundaries so birds never trespass.
3. **Quiet enough for the setting:** Managed without roosters, with realistic expectations for hen noise.
4. **Legally placed:** Situated well within required setbacks and property limits.
5. **Odor-controlled:** Kept dry and well-ventilated to prevent ammonia buildup.
6. **Visually reasonable:** Free of messy tarps, scattered equipment, and trash.
7. **Responsive:** Guided by an owner who listens to concerns and takes rapid corrective action.

What neighbors usually notice

To prevent conflict proactively, you must step into your neighbor's shoes. Ask yourself: What would my closest neighbor see, hear, or smell from my planned coop location? When backyard chicken keeping goes wrong in the suburbs, neighbors notice a highly predictable set of issues. Relying on good luck is not a strategy. Instead, use the following matrix to understand the neighborhood impact of your flock, the root causes of those impacts, and the specific actions required to prevent and fix them.

Odor

- Primary Cause: Damp manure buildup; poor ventilation; overflowing compost.
- Prevention Strategy: Keep bedding dry; remove waste regularly; design run for drainage; turn compost weekly.
- Emergency Fix: Rake out and remove all wet bedding immediately; replace with a thick layer of dry wood chips.

Flies

- Primary Cause: Poor moisture management; spilled feed mixing with mud; unmanaged droppings.
- Prevention Strategy: Keep run surface bone-dry; clean up food spills daily; use tight, enclosed compost bins.
- Emergency Fix: Shovel out damp feed paste; then use fly traps only as one part of an IPM plan after fixing manure, moisture, spilled feed, garbage, and drainage sources. Choose traps labeled or designed for use around animals or poultry, avoid insecticide baits unless the label allows that site, and place attractant or odor traps where they pull flies away from the coop, doors, patios, and neighbor-use areas while remaining out of reach of chickens, pets, children, and wildlife.

Rodents

- Primary Cause: Accessible feed; spilled scratch grains; overnight food scraps; insecure storage.
- Prevention Strategy: Lock all feed in heavy-duty galvanized metal bins; remove uneaten scraps before dusk.
- Emergency Fix: Secure all food sources immediately; consult Extension-backed pest control guidance for safe trapping.

Loose Hens

- Primary Cause: Gaps in fencing; unfenced free-ranging; unsecured gates.
- Prevention Strategy: Inspect fences regularly; use secure, locking gates; keep hens in a fully enclosed run.
- Emergency Fix: Return hens to the secure run immediately; patch fence gaps before their next turnout.

Noise

- Primary Cause: The daily "egg song"; squabbling over nest boxes; accidental roosters.
- Prevention Strategy: Provide adequate space and sufficient nest boxes; do not promise total silence.
- Emergency Fix: Block visual sightlines to trigger-noises (like dogs); immediately isolate and rehome any accidental roosters.

Visual Clutter

- Primary Cause: Coop crowding the property line; messy feed and tool storage; temporary tarps.
- Prevention Strategy: Respect all setbacks; keep supplies in sealed bins out of view; use solid visual fence barriers.
- Emergency Fix: Clean up surrounding trash; store ripped feed bags in bins; remove deteriorating plastic tarps.

The neighbor-friendly promise

Shifting from defensiveness to stewardship means making a neighbor-friendly promise you can actually keep. This is a set of internal standards you commit to before a single bird arrives.

Rather than relying on vague goals, use this concrete checklist to set your neighborhood standards:

- **I will ensure my flock is legal** under every municipality, county, HOA, or lease rule that applies to my property.
- **I will keep my hens securely contained** in a run so they never become a nuisance in someone else's yard.
- **I will store feed in pest-resistant containers** and clean up major spills to reduce the attraction of rodents.
- **I will manage, compost, or safely dispose of waste** according to local rules and my pest-control plan so odor stays controlled.
- **I will place the setup thoughtfully**, factoring in sightlines, drainage, property setbacks, and my neighbors' daily experience.
- **I will respond promptly and calmly** to any neighbor complaints, taking immediate action to fix the root cause.

Reference **Neighbor-readiness checklist in Appendix B.**

The trap of "egg diplomacy"

A common trap for eager beginners is the myth of egg diplomacy. The thinking goes: "If I just bring my neighbors a free carton of fresh eggs every few weeks, they will look the other way if things get a little noisy or smelly."

This is a dangerous miscalculation. A carton of fresh eggs is a lovely gesture, but it is not currency. You cannot use eggs to purchase the right to create a neighborhood nuisance. A dozen eggs will never make up for a swarm of flies ruining a family barbecue next door, nor will it stop rats from migrating under a shared fence. Give your neighbors eggs because you want to be generous, but never use them as a substitute for strict sanitation and secure boundaries.

When and how to talk to neighbors

A critical part of suburban chicken keeping is knowing when and how to talk to the people next door.

Do not ask your neighbors for permission before you know the rules. Your first job is to verify what is legally allowed on your property. You do not want to ask your neighbors to arbitrate local laws or solve your zoning mysteries.

Once you know your plan is legal, but before you break ground or purchase supplies, have a brief, non-defensive conversation with your immediate neighbors. Keep the conversation informational. The goal is to set expectations, not to ask for a vote.

Use this polite neighbor conversation script:

"Hi there, I wanted to give you a quick heads-up. I have been researching backyard chickens, checking the city codes, and I am planning to set up a small flock of four hens next month. I want to assure you I have a strict plan for keeping the feed secure in metal bins, managing the waste so odor stays controlled, and keeping the birds completely contained in their run. The coop will be placed on the far side of my garage, so you shouldn't see much of anything. I just wanted to let you know beforehand, and if you ever notice an issue, please tell me directly so I can fix it right away."

When you initiate this conversation, some neighbors will be thrilled. Others will immediately look worried. If a neighbor expresses concern, listen first. Do not interrupt to aggressively defend your property rights.

Instead, ask what specific issue worries them. Are they afraid of waking up to a rooster crowing? Acknowledge their concern, explain that you plan to keep only hens, and assure them you have a strict plan to rehome any accidental roosters. Are they terrified of rats because they had a pest problem a few years ago? Explain the exact galvanized steel bins you bought to lock up the feed.

At the same time, **do not overpromise**. If you promise your neighbor they will absolutely never hear a single cluck, smell a single dropping, or see a single fly, you have set yourself up to fail on day one. Instead, promise active management. Promise that the flock will be cared for properly and that you will be a responsive neighbor if an issue arises.

Complaint response plan

Even with the best preparation, a neighbor might eventually complain. Complaints do not automatically mean you have failed as a flock owner. They are usually just routine maintenance alerts indicating that part of your backyard system needs adjustment.

When a neighbor brings an issue to your attention, rely on a structured response plan rather than reacting emotionally. First, document the complaint. Second, inspect the relevant issue. Third, take a corrective action. Finally, follow up politely to confirm the issue is resolved.

It is vital to use language that de-escalates the situation **without making unnecessary statements about fault or legal conclusions**. If a neighbor complains, do not reply with texts or emails saying, "I am so sorry my flock is causing a nuisance," or "I completely failed at keeping the flies away." If the neighbor escalates the issue to a homeowner's association, a landlord, or city code enforcement, those casual apologies may be quoted back to you and, depending on local law and the forum, could be characterized as admissions or as evidence that undercuts your position. For a serious or escalating dispute, get local legal advice rather than relying on a script.

Instead, stick to process-oriented, factual responses. Acknowledge the communication, state that you are looking into it, fix the problem, and report back.

Complaint-response scripts:

- **Initial acknowledgment:** "Thanks for reaching out and letting me know. I am heading out to the yard right now to take a look at the fence line."
- **Follow-up after action:** "Thanks again for the heads-up. I found where some water spilled near the fence and attracted flies. I have cleaned up the area, put down fresh dry wood chips, and moved the waterer. Please let me know if you see them over there tomorrow."

Imagine your neighbor texts you on a warm Tuesday afternoon. They are polite but annoyed, stating there are suddenly dozens of flies swarming near the shared fence.

Your first step is to listen and acknowledge using the script above. You walk out to the coop and inspect the area. You notice that your hens have been enthusiastically kicking their scratch grains out of the feeder. The spilled feed has mixed with damp soil near the edge of the run, creating a fermented, messy paste right near the property line.

You take immediate corrective action. You rake up the wet feed and damp bedding, placing it into a secure, pest-resistant compost bin. You lay down a thick layer of dry, fresh wood chips, and you adjust the height of the feeder so the hens cannot easily kick the food out. An hour later, you send the factual follow-up text confirming the area has been cleared.

By making and keeping this neighbor-friendly promise, you remove the daily anxiety of waiting for an angry knock on the door. You know exactly what your standards are and how to maintain them. With the local rules verified and your neighborhood boundaries respected, you are finally ready to look at what this well-managed suburban flock will actually cost to build.

Chapter 4: Budget for the flock you will actually own

You have probably heard someone say they want to get backyard chickens for the free eggs. It is a lovely idea, picturing yourself walking out the back door and returning to the kitchen with breakfast.

But experienced suburban chicken keepers know a different reality. The eggs sitting on the counter are not a financial windfall. Behind every carton of backyard eggs is a hidden, mostly permanent support system. That system includes a secure coop, a spacious run, weatherproof feed storage, bulk bedding, predator-resistant hardware, health resources, and daily time.

Backyard chickens produce useful compost material and are entertaining to keep, but they should never be sold to beginners as guaranteed grocery savings. Treat chickens like a vending machine, and you will be deeply disappointed by the return on your investment. To build a sensible, safe setup, you must stop guessing and budget for the flock you will actually own.

The first-year reality

Your setup costs happen long before the first egg arrives. If you buy day-old chicks, you will feed, house, and clean up after them for up to six months before a single pullet lays an egg. Once they start, production varies naturally, slowing down during winter or when hens molt.

Meanwhile, recurring costs never take a break. Feed, bedding, coop repairs, and health needs continue year-round. Furthermore, flock size directly changes your budget. The difference between four hens and eight hens is not just extra feed; it requires a larger run, more bedding, bulkier storage, extra waste-handling capacity, and more extensive vacation care.

Because local climates, feed prices, and housing needs vary widely, national averages are mostly useless. You need to price your specific local reality.

The startup and ongoing categories

Organize your initial spreadsheet into concrete categories. Your one-time setup expenses will arrive before you ever bring a bird onto your property.

Startup categories:

- **Coop and run:** The physical structure, whether a DIY build, a premium shed conversion, or a heavily modified prefab kit.
- **Predator reinforcement:** Heavy-duty, welded-wire hardware cloth, secure locking carabiners, heavy-duty hinges, and trenching materials. Standard chicken wire is not predator-resistant.
- **Feeders and waterers:** Elevated or suspended systems to deter pests and prevent spills.

- **Bedding:** Initial bulk pine shavings, hemp, or construction sand.
- **Initial feed and supplements:** Layer feed, plus supplemental calcium (oyster shell) and grit.
- **Brooder supplies (if raising chicks):** A secure box, safe heat source, chick starter feed, and shallow waterers.
- **Permits and approvals:** City permit fees, inspection costs, or HOA application fees.
- **Secure storage:** Galvanized metal trash cans with tight-fitting lids for feed.
- **Cleaning tools:** Dedicated rake, flat scraper, wheelbarrow, and heavy-duty buckets.
- **Weather gear:** Tarps, shade sails, or windbreaks suited to your climate.

Ongoing categories:

- **Feed and supplements:** Ongoing monthly replacement.
- **Bedding replacement:** Regular replenishment to keep the coop dry and manage waste.
- **Hardware and repairs:** Replacing rusted locks, patching roof leaks, or adding winterization materials.
- **Pest prevention:** Replacing worn weatherstripping or restocking traps.
- **Seasonal needs:** Electrolytes for summer heat, or extra pine shavings for winter.
- **Veterinary, diagnostic, or emergency care:** Mail-in fecal tests, diagnostic consultations, or basic first-aid supplies.
- **Vacation care:** Paying a reliable sitter to check supplies, collect eggs, and secure the flock at night.

The price-per-pound feed prompt

Feed will be your largest ongoing expense. Many beginners make the mistake of comparing feed bags strictly by the sticker price. A $25 bag of feed might seem like a worse deal than a $22 bag, but if the first bag holds 50 pounds and the second holds only 30 pounds, the cheaper bag is actually much more expensive over time.

Prompt: Visit your local farm or hardware store. Find three brands of layer feed. Divide the total price by the weight of the bag to find the price per pound. Factor your local tax into the final cost. Use this baseline per-pound calculation to estimate your ongoing monthly feed budget.

Three first-year budget scenarios

When evaluating initial startup costs, divide your spending into three tiers. Do not spend a single dollar on decorative upgrades, like custom signs or matching curtains, until the minimum safe setup is fully funded.

To help you visualize the required investment, review the following three budget scenarios. These tables assume a standard four-hen suburban flock.

Note: The estimated ranges below are structural guidelines requiring local verification. Prices exclude the cost of the birds themselves and fluctuate based on regional lumber costs, inflation, and local supply.

Scenario	Profile	Estimated Range (First Year)	Assumptions
1. Minimum Functional	DIY from plans using basic materials. Secure, legal, and safe, but no frills. Requires physical labor to build.	$800 – $1,200	Uses standard framing lumber, heavy-duty hardware cloth, and basic roofing. Includes basic metal storage bins, standard feeders, and minimum required square footage. Assumes owner does all labor.
2. Prefab Retrofit	Buying a standard online coop kit but spending the necessary money to reinforce it against predators and weather.	$1,200 – $1,800	Includes the base cost of the kit, plus required upgrades: rolls of hardware cloth to replace cheap wire, predator apron materials, extra lumber for run expansion, better latches, and weather-sealing paint/roofing.
3. Comfortable Beginner	A premium shed conversion or heavy-duty walk-in setup. Built for owner convenience, standing height, and long-term durability.	$2,000 – $3,500+	Includes a sturdy shed base, custom heavy-duty run framing, covered roof over the run to prevent mud, premium easy-to-clean feeders, and bulk feed storage. Saves significant time on daily chores.

Budgeting for the emergency fund

Suburban chicken keeping requires an emergency contingency fund because the real world will inevitably disrupt your spreadsheet. It is entirely realistic to face one or more of the following events in your first year:

- **The predator breach:** A raccoon bends the cheap wire you decided not to upgrade, or a neighborhood dog digs under the run. You suddenly need $100 for heavy-duty hardware cloth, a box of screws, and a predator apron to fix the perimeter before nightfall.
- **The health crisis:** One of your hens develops a severe limp, or you suspect a contagious disease is moving through the flock. You need $150 to $300 for a local avian veterinarian exam, a diagnostic mail-in test, or approved emergency first-aid supplies from your farm store.
- **The structural failure:** An intense winter storm collapses your run tarp, or heavy spring rains flood the coop floor. You must immediately purchase $75 worth of construction sand, new gravel for drainage, or heavy-duty corrugated roofing panels.

Expect the unexpected. Keep a dedicated financial reserve, ideally $200 to $400, specifically earmarked for the flock.

The time-cost table

Dollars are not the only currency you spend on your flock. Time is a finite resource that must be budgeted just as strictly. You cannot leave town for a long weekend and hope the birds manage themselves.

Before committing, review the standard time requirements for a suburban flock.

Frequency	Standard Tasks	Estimated Time Commitment
Daily	Opening the coop, checking the flock for injuries/illness, refilling waterers, topping up feed, collecting eggs, locking up securely at dusk.	10 to 15 minutes per day.
Weekly	Scrubbing waterers to prevent algae, scraping the dropping boards, adding fresh bedding as needed, taking out compost or waste.	30 to 45 minutes per week.
Monthly	Deep cleaning the coop, checking hardware and latches for rust or predator damage, hauling heavy bags of feed and bedding from the store, updating records.	1 to 2 hours per month.
Seasonal	Winterizing the coop, adding weather breaks, thoroughly power-washing and sanitizing, preparing for summer heat protocols, evaluating flock health.	3 to 4 hours per season.

Managing neighborhood relationships takes time as well. Whether that means chatting with the neighbor over the fence, answering questions about noise, or keeping your local rule-compliance records properly filed, you must budget time to be an ambassador for your flock.

Your next steps

Before deciding how many chickens to bring home, price out the realities of your specific neighborhood. Stop guessing and start building a verifiable plan.

Asset Check: Use Appendix H: First-year budget worksheet and Appendix I: Supplies and shopping list to organize your local prices.

Exercise: Take your worksheets to your closest farm and hardware stores. Find the exact prices for a 40-pound bag of layer feed, a bale of pine shavings, galvanized metal trash cans, hardware cloth, and durable feeders. Write them down.

By replacing vague financial hopes with clear categories, verifiable local prices, and realistic time expectations, you are investing in a sensible infrastructure. Once you understand exactly what it costs to support a flock securely, it is time to look out your back window. Your next step is figuring out exactly where this new daily operation is going to live.

Part 2: Build the suburban chicken setup

You have run the numbers and committed to the process. Now, you have to build something physical. Walk out your back door and stand on the grass. Look around your property. Somewhere in this enclosed, finite space, you need to establish a permanent living area for animals that function more like livestock than traditional pets. They are not outdoor cats that roam freely and return to the porch when they are hungry. They are not wild songbirds passing through to visit a feeder. They are ground-dwelling, scratching, vulnerable creatures that will rely entirely on the physical environment you construct for them. While some suburban keepers attempt to rotate small, movable enclosures, most rely on a fixed, permanent setup. In a residential yard, getting that physical location right on the first try is critical.

It is incredibly tempting to jump straight into the visually rewarding part of the project. You probably want to go online, buy the most picturesque wooden coop with a scalloped roof, assemble it on a Saturday afternoon, and fill it with birds by Sunday. But a residential lot is an unforgiving place for rushed construction, and a poorly chosen setup will fight you every single day.

On a large rural property, if a farmer places a coop in a low spot and the ground turns muddy, they can simply hook the structure up to a tractor and drag it to a drier patch of pasture. In a fenced suburban yard, your options are permanently restricted. If you place your coop in a soggy patch under a roof downspout, your chickens will stand in a foul-smelling swamp, and you will be stuck trenching water away from it by hand. If you build right against a back fence line without thinking about maintenance access, you will quickly discover that you have no room to swing a shovel, no clearance to operate a wheelbarrow, and no way to reach the back wall to repair a loose board.

Your setup dictates your joy. The physical infrastructure you build over the next few weeks will determine whether keeping chickens is a manageable morning routine or a grueling, filthy chore. Consider hauling a heavy, heated waterer in February. If your coop is tucked at the very back of a sloping lot, you will be slipping on ice every morning. If the access doors are too small or positioned awkwardly, you will find yourself crawling on your hands and knees into soiled bedding just to reach a stubborn corner. A thoughtfully designed setup prioritizes human ergonomics, meaning you can easily reach the nest boxes, refill water, and scoop bedding without hitting your head, straining your back, or wading through muck. A poor setup means you will dread going out to the low corner by the shed when it is freezing and raining. Neglected chores quickly lead to damp bedding, pest infestations, and sick birds.

Over the upcoming chapters, we will cover the physical workflow required to establish your yard, choose your coop, and secure your run. But before you buy materials, you need to map your landscape. Coop placement is a delicate physical balancing act. It requires evaluating drainage and water pooling after heavy rain. It requires understanding sun, shade, wind, and seasonal exposure so your birds do not overheat in July or freeze in January. It also requires charting a daily access route that makes feeding, watering, cleaning, and hauling heavy yard waste practical in all four seasons.

Once the location is mapped, the housing itself comes into focus. We will build your setup around a simple rule: the coop is the bedroom; the run is the living room. Beginners routinely spend their entire budget on a beautiful, tiny coop while completely ignoring the outdoor run. We are going to flip that priority. Hens spend their waking hours scratching, dust bathing, and moving around. A secure, draft-free, well-ventilated coop is essential for sleeping safely and laying eggs, but a spacious, covered run gives your flock the room they desperately need to express their natural behaviors without turning your lawn into a barren, eroded wasteland.

Simultaneously, you are building a defensive perimeter. Beginners often assume that a standard six-foot wooden privacy fence will keep their birds safe. The reality is that residential greenbelts are thriving, high-traffic habitats for hungry wildlife. Raccoons possess incredible dexterity, capable of unlatching simple locks and reaching through thin wire to grab whatever is inside. Hawks sit quietly on telephone poles and neighborhood oak trees, watching the grass. Foxes adapt remarkably well to subdivisions. Furthermore, one of the most common threats to a backyard flock is a domestic dog with a high prey drive. Because no setup can be guaranteed completely predator-proof or dog-proof, your coop and run must be designed as the actual physical barrier. We will transition away from flimsy materials like traditional chicken wire, which only keeps chickens contained, and focus entirely on structural hardware cloth, heavy-duty latches, inspectable mesh, and trenching techniques that offer genuine predator resistance.

Finally, we will design your physical layout to address the unglamorous side of suburban chicken keeping: waste management. Chickens produce a significant volume of manure every single day. Where there is poorly managed manure, there is inevitable odor. Spilled feed left on the ground increases the risk of rodents, and persistent moisture attracts flies. You must have a physical plan for where the soiled bedding will go once you scoop it out. Will you build a dedicated compost bin, or will you bag it for yard-waste collection? In a densely packed neighborhood, waiting until a rat shows up or the smell becomes overwhelming means you are already losing the battle. We will build odor and pest prevention directly into your physical infrastructure from day one. Secure feed storage, smart drainage, and dry bedding systems must be planned before the birds arrive, ensuring your yard remains clean and your flock stays healthy.

The spatial decisions you make right now form the permanent foundation of your suburban flock. Take your time. Do not rush to the farm store just yet. Grab a tape measure, pull on a pair of boots, and step outside. Your physical build starts with the ground beneath your feet.

Chapter 5: Map your yard for setbacks, drainage, shade, and access

MOST SUBURBAN CHICKEN KEEPERS make the exact same mistake when choosing where to put their first coop. They walk out the back door, scan the yard, and immediately spot the "empty corner."

Often situated at the far back property line, perhaps tucked behind the garage or nestled under a large tree, this spot is currently doing nothing. You do not have to look at it from the kitchen window, so it seems like the ideal place to put a box of chickens. You might even be thinking right now that the back corner is best precisely because it is out of sight.

But placing a chicken coop is not like setting down a birdbath or a decorative garden bench. A residential lot is an unforgiving place for rushed construction, and a poorly chosen setup will fight you every single day. Choosing a location is more like deciding where to permanently install a heavy trash can, a daily-use garden hose, and a messy pet shelter all at once. If you put that setup in a remote corner, you will regret the decision the first time you haul two sloshing water buckets through January snow. If you choose a low spot, you will spend half the year battling foul-smelling mud. And if you accidentally place it right against the property line, you might violate a local setback rule and face an expensive tear-down order.

The best coop location is chosen on paper before anything is built, bought, or delivered. Your goal right now is to turn an abstract idea of your yard into a layered, practical planning workflow.

Layer 1: Build the base map

To get started, you need paper, a pencil, and a clipboard.

Draw the basic outline of your property lines. Do not just trace your existing fence. Fences are frequently built a foot or more inside or outside actual property boundaries. If you are uncertain where your legal property lines are, verify them using your property pins or a local plot plan before you assume you have room for a coop.

Inside those legal boundaries, sketch the footprint of your house. Mark the back doors, gates, fences, patios, decks, and any existing sheds. Next, rough in the major trees, children's play areas, vegetable gardens, and zones where your dog currently runs or uses the bathroom. Do not worry about drawing to perfect architectural scale; you just need a functional bird's-eye view. Finish by noting the sides of your yard that face your neighbors or the street.

Pull up Appendix D Yard Map and Coop Placement during this chapter.

People often assume they can simply move the coop later if they need to. In reality, even a small suburban coop designed for four hens is incredibly heavy once loaded with deep bedding, feeders, and roofing materials. When you attach a secure, predator-resistant run, the structure becomes inconvenient and expensive to undo. You are making a long-term real estate decision for your yard.

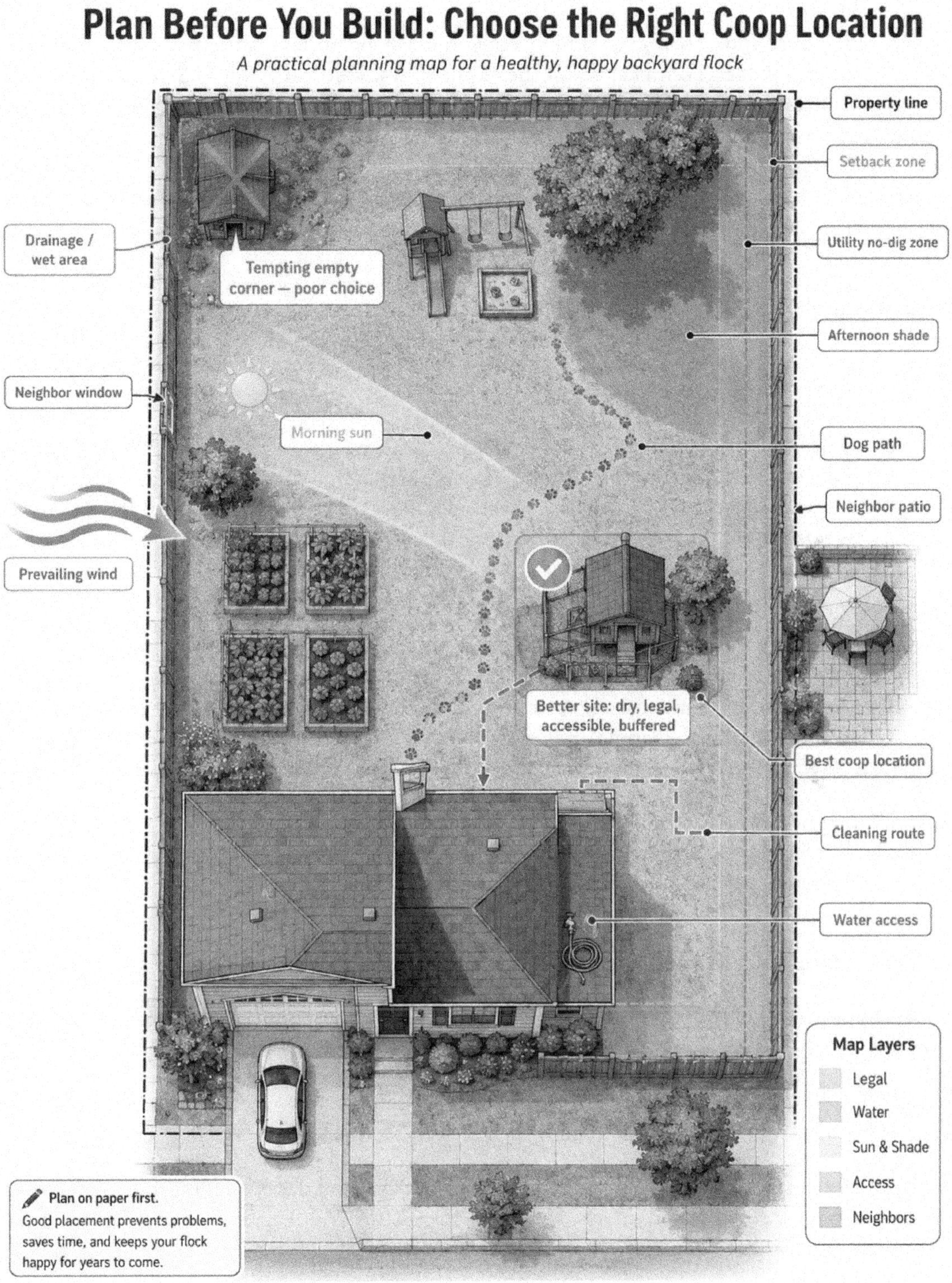

Layer 2: The legal, setback, and utility overlay

With the physical objects drawn, you must map the invisible legal boundaries that dictate where structures can and cannot go. Grab the rules you verified in Chapter 2.

If your local city or county requires a specific setback, whatever distance your ordinance specifies from property lines, neighbor dwellings, or streets, draw those exclusion zones on your sketch. Shade them out heavily with your pencil. You absolutely cannot put a coop in those spaces, no matter how convenient they seem.

Next, shade out any known utility easements, septic fields, or municipal drainage ditches.

Finally, mark your utility hazard zones. Before trenching, augering, driving posts, or otherwise disturbing soil, contact 811 or your state/local one-call utility-locate center and wait the required time before digging. U.S. requirements, notice periods, and exemptions are set by state law, so check your state 811 center's rules. You might think installing a chicken coop does not require digging, but proper predator resistance involves burying hardware cloth aprons in a trench or driving posts for a run. You do not want to strike a shallow internet cable, a water line, or a pressurized gas line while building a chicken run. Note on your map where utilities enter your house, and treat those paths as no-build zones until a professional has marked them. Remember that one-call locating may not cover private lines on your property, which may require additional locating.

Layer 3: The drainage and stormwater overlay

Next, map your yard for water. Mud is the enemy of backyard flock health. Wet, manure-heavy ground harbors bacteria, increases odor, invites flies, and causes serious foot and feather issues for your birds.

Consider the classic case of the "pretty corner." A new keeper places a freshly painted coop in the far back corner under a weeping willow. During the dry weeks of July, it is a perfect pastoral scene. But when autumn rains arrive, they discover the entire yard slopes gently toward that corner, and the soil there stays perpetually damp. Suddenly, the pristine chicken run is a swamp of wet pine shavings and standing water. The "pretty corner" fails because it was evaluated on aesthetics rather than drainage.

Identify the low spots on your sketch. Note where gutters and roof runoff empty into the yard. Highlight areas that stay spongy for days after a storm, and pay attention to slopes that send rainwater washing across the lawn.

You must also account for stormwater runoff leaving your property. [VERIFY: Extension drainage best practices]. Heavy rain washing through a chicken run will carry manure and excess nutrients with it. You cannot place a coop where stormwater will flush chicken waste directly onto a neighbor's patio, into a shared swale, or toward a nearby creek or storm drain.

Action Prompt: Walk your yard after a heavy rain before choosing your coop spot. Put on your boots, go outside, and physically look for standing water, spongy soil, and active runoff channels. Check for muddy walking routes and wet gates. If water pools in an area, do not put chickens there. Cross those muddy zones off your map.

Layer 4: The sun, shade, wind, and canopy overlay

Weather behaves differently in various parts of your yard. It is easy to assume that shade is always best for animals, but chickens require a balance.

For many suburban flocks, extreme summer heat is deadlier and much harder to manage than winter cold. A poorly ventilated wooden coop sitting in direct afternoon sun can quickly turn into an oven.

Observe your yard's microclimates. Morning sun is highly beneficial. It dries out damp bedding, burns off overnight dew, and warms the flock after a cold night. However, intense afternoon sun in July and August is a hazard. A location that receives gentle morning sunlight but is shielded by deep afternoon shade is often ideal. Sketch these shifting sun and shade zones on your map.

Next, consider the wind. If you live in an area with harsh winters, notice where storms typically hit your yard to determine the prevailing winter wind direction (often north or northwest in many regions). You do not want the main doors and ventilation windows facing directly into freezing, driving sleet. Tucking the coop against a solid fence or a sturdy garage wall can provide a windbreak, provided it sits outside your legal setbacks and leaves enough room for airflow and maintenance.

Finally, look up. If you plan to put the coop under a mature tree for that valuable afternoon shade, evaluate the health of the branches overhead. A heavy dead limb falling during a summer thunderstorm or a winter ice storm will easily crush a standard backyard coop. Cross out any zones directly beneath dead or dying tree canopy.

Layer 5: The access and workflow overlay

Now, draw a line representing the path you will walk from your back door to the proposed coop site. You will walk this route at least twice a day, every day, for years. You will walk it when you are running late for work, when it is raining sideways, and when you step out in your slippers to grab eggs on a Sunday morning.

Ask yourself a few practical questions to test this workflow:

- **Water access:** Chickens drink a surprisingly large amount of water. If your garden hose does not reach the coop, you will be carrying heavy buckets across the lawn. In winter, when the hose is frozen and put away, you will definitely be carrying buckets. The closer the coop is to a reliable year-round water source, the happier you will be.

- **Feed storage:** Where will you store the heavy bags of feed, and how far will you have to carry them? If your feed is secured in the garage but the coop is on the opposite corner of the property, you are signing up for a long haul.

- **Cleaning route:** Every few weeks, you will need to wheel a barrow full of soiled bedding to a compost bin or disposal area. Make sure your path is wide enough for a wheelbarrow and does not require hauling dirty, odor-heavy bedding across a clean outdoor dining patio.

The final workflow layer involves the people and animals who share your space. It is a common misconception that if a coop simply looks nice, placement does not matter. But even a beautifully kept coop will generate occasional noise, a few flies, and dust from the birds taking dirt baths.

Highlight the neighborhood conflict zones on your drawing. Where are your neighbor's bedroom windows? Where are their patios, barbecue grills, or pools? If you place the coop right next to their outdoor dining table, you risk triggering a nuisance complaint. Try to use your garage, your house, or dense landscaping as a buffer. Because appearance matters in many suburban areas, flag the spaces that are highly exposed to public view from the street or an HOA-monitored sidewalk.

Do the same for your own household. Keep the coop away from children's swingsets to avoid trip hazards and unsupervised interaction. Keep the coop away from the fence line where your dog naturally paces, preventing constant barrier-frustration barking and chronic stress for the birds.

The coop-site scoring matrix

By now, your yard map should look entirely different than it did when you started. Large sections will be heavily crossed out because they violate a setback, sit in a muddy depression, share an underground utility line, or sit uncomfortably close to a neighbor's open window.

Look at the open spaces that remain. Choose two or three candidate locations that balance good drainage, reasonable distance for daily access, and respectful neighborhood boundaries.

Take your map and physically stand in your candidate spots. Photograph each location from multiple angles to reference later. Then, use the following scoring matrix (also found in the companion PDF) to evaluate each site. Score each category on a simple scale of 1 to 5, where 1 is poor/difficult and 5 is excellent/easy.

Candidate Site Scoring Matrix:

- **Legal & Utility Clearances (1-5):** Is it well outside the setback zone, away from utility lines, and clear of HOA sightlines?

- **Drainage & Footing (1-5):** Is the ground high, firm, and safe from heavy stormwater runoff?

- **Sun & Shade Balance (1-5):** Does it get morning sun to dry out, with critical shade protection from the harsh afternoon heat?

- **Weather & Wind Protection (1-5):** Is it shielded from the prevailing winter wind, and free from dangerous overhead branches?
- **Water & Feed Access (1-5):** How easy is the winter water haul? Can you reach it with a hose in the summer?
- **Cleaning Route (1-5):** Is there a clear, flat path for a wheelbarrow between the coop and the compost/disposal area?
- **Household & Neighbor Harmony (1-5):** Is it buffered from neighbor patios, dog paths, and children's play equipment?

Tally the scores to see which locations pull ahead, but do not commit to a final spot just yet. You have mapped the raw land, but the structure you place there matters just as much. You need to understand how the physical coop itself will sit on the ground, how its doors will swing, and how you will clean it out. Once you grasp the mechanics of a well-placed structure in Chapter 6, your winning yard location will become obvious.

Use the space below to start making some notes and jotting down your thoughts.

Chapter 6: Place the coop for peace, cleanliness, and convenience

In the previous chapter, you mapped your yard to identify legal setbacks, utility lines, and drainage zones. You eliminated the illegal, unsafe, and flooded areas. Now, you must translate that two-dimensional map into a three-dimensional physical workflow.

Your goal is to choose a location that is, above all else, convenient enough to maintain. When backyard routines become physically burdensome, they break down. When routines break down, coops get dirty, predators find vulnerabilities, and neighbors complain. Pull up Appendix E – Coop and Run Checklist with this chapter.

The 360-degree footprint

A coop is not a solid block; it is an active mechanical structure with doors, vents, ramps, and moving parts. You cannot simply drop it onto a patch of grass. You must account for the 360-degree footprint required to operate and maintain the building.

Repair clearance: Never push a coop flush against a property fence or a garage wall. Wood rots, hardware cloth sags, and hinges rust. Eventually, you will need to squeeze behind the structure with a drill, a hammer, or a roll of heavy wire mesh. If you do not leave at least three feet of physical clearance on all four sides, basic maintenance becomes impossible without physically dismantling or dragging the entire coop.

The clean-out side: Every well-designed coop has a primary clean-out door, usually a large set of double doors or a drop-down wall that allows you to sweep soiled bedding directly out of the structure. Identify which direction this door faces. You must have enough open space on the clean-out side to park a wheelbarrow directly under the door, stand beside it, and comfortably swing a shovel. If the clean-out side faces a retaining wall or a dense hedge, you will find yourself carrying dirty manure by the handful.

Door swing and gate width: Before finalizing a spot, simulate opening every door. A human-sized run door that swings outward into a narrow gap will trap you against a fence. A solid clean-out door that swings backward into a tree trunk will only open halfway. Pay equal attention to gate width. The path leading to your coop, and the run door itself, must be wide enough to accommodate a fully loaded standard wheelbarrow. If your wheelbarrow cannot clear the run gate, you will spend your weekends hauling bedding in awkward, heavy buckets.

Nest-box access: You will collect eggs daily. The exterior nest-box lid must be easy to reach, free from overhead obstructions, and clear of ground-level trip hazards. If you have to step over a drainage ditch or duck beneath low-hanging branches just to gather eggs, you have chosen the wrong orientation.

Run attachment: Prefabricated and custom coops alike usually feature a dedicated exit door for the chickens to enter their run. Consider how the enclosed run attaches to the main housing. The ground beneath the run must be relatively flat to prevent wide gaps from forming between the bottom frame and the soil. Placing the setup on a steep slope requires extensive leveling, trenching, and foundation work to keep the run secure.

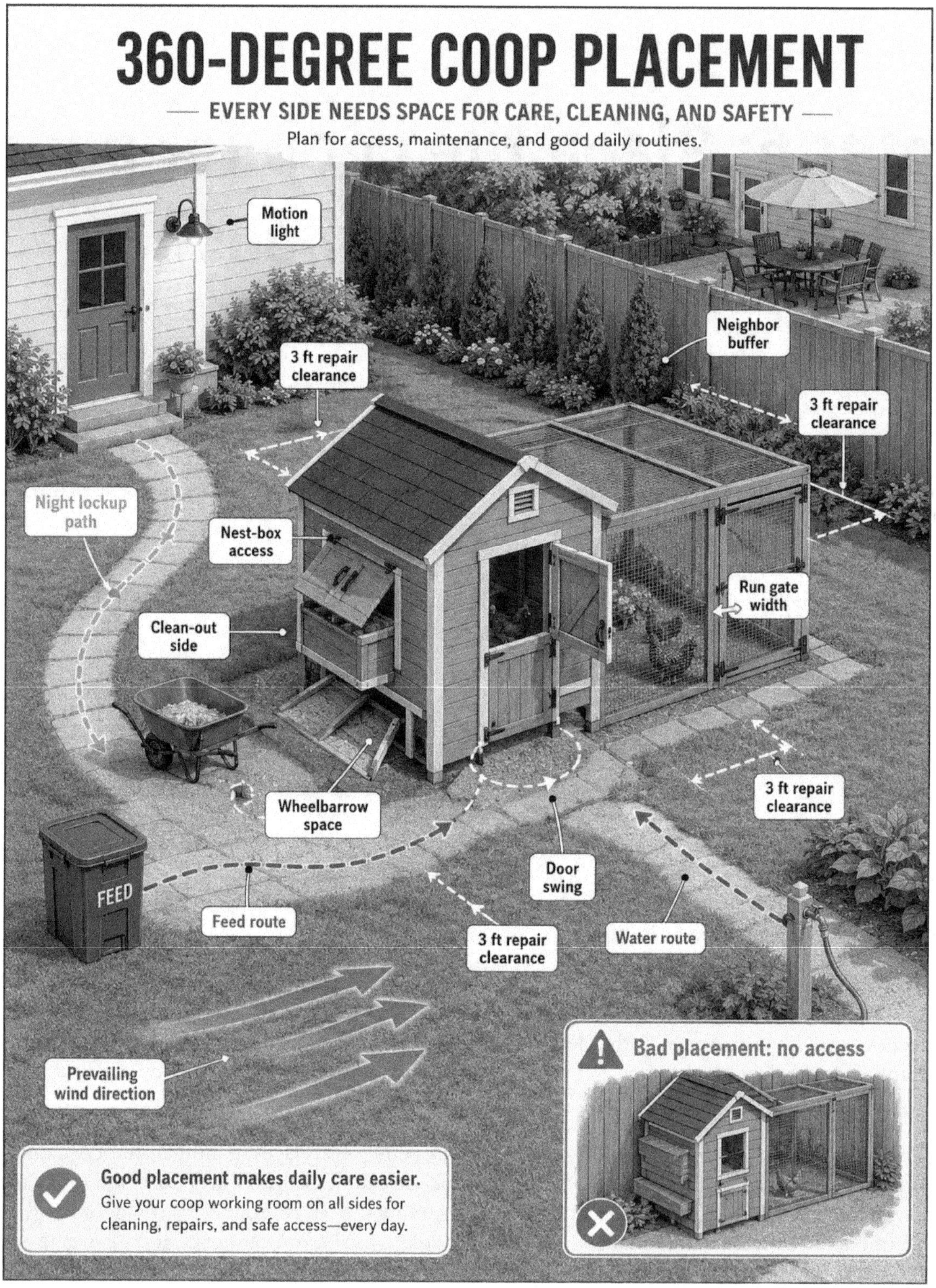

The daily workflow: feed, water, and lighting

Try this practical exercise. Grab a heavy bucket. Fill it with water or a few large rocks to simulate the weight. Now, walk the exact route from your kitchen door to your planned coop location.

Note every obstacle. Do you have to unlatch a sticky side gate? Step over a garden hose? Navigate around the spot where the snowplow piles winter drifts? You will make this walk hundreds of times a year, often carrying fifty-pound feed bags, heavy bales of pine shavings, or delicate cartons of fresh eggs.

Feed and water workflow: Water logistics dictate coop placement more than almost any other factor. In the summer, you need an easy route to drag a garden hose for scrubbing and refilling waterers. In the winter, when outdoor hoses freeze, you will be carrying liquid water from the house to the coop. Choose a location that minimizes this carry distance. Similarly, map out your feed storage workflow. Feed must be kept in secure, rodent-resistant metal bins, ideally in a garage or tightly sealed shed. Placing the coop on the opposite side of the property from your feed storage means hauling heavy scoops of grain across the yard in all weather conditions.

Path lighting and night checks: For most suburban flocks, build your workflow around securing the birds at dusk in a predator-resistant coop or comparably secure enclosed setup appropriate to your property and local predator pressure. In the winter, the sun will set long before you finish dinner. You will be walking out to the coop in the pitch black. The route must be free of trip hazards and well-lit. If the path requires navigating uneven paving stones, discarded children's toys, or complex garden features in the dark, you will dread the nightly lockup. A straight, predictable path illuminated by a motion-sensor light should be treated as a core safety requirement for suburban flock keeping.

Odor paths, wind, and neighbors

"If my setup is legally placed inside the setback lines, my neighbors shouldn't care where it goes."

This is a common reaction among beginners, but it is a risky way to build long-term neighborhood peace. Legal compliance does not automatically equal neighborly consideration. You must actively manage how your flock's physical location impacts the people living next door.

Odor path and prevailing wind: Even a well-managed coop generates dust, dander, and occasional odors. Wind carries these directly across property lines. Research the prevailing wind direction in your area and map the likely odor path. If the prevailing summer breeze blows directly from your planned coop site onto your neighbor's primary outdoor dining patio, move the coop.

Neighbor sightlines: Stand at your neighbor's likely viewpoint and look closely at your planned location. What do they see from their kitchen window or deck? An orderly, well-maintained coop can be a charming backyard feature, but a setup surrounded by stray tools, empty feed bags, and a towering pile of dirty bedding will draw complaints. If your best placement option remains visible to neighbors, plan to install a wooden lattice panel, a privacy screen, or simple evergreen landscaping to soften the view.

Household boundaries and helper access

Your chickens share the property with your household. The coop must not clash with how your family and your pets already use the yard.

Children and dogs: Establish a hard, physical boundary between where children play and where chickens live. Avoid placing the setup bordering swing sets, sandboxes, or primary lawn play zones. Chickens kick debris out of their runs, and their living area must be treated as a strict hygiene boundary. Furthermore, separate the coop from established dog paths. Even a friendly family dog with a low prey drive will cause immense stress to a flock if it runs energetic patrol laps directly against the wire. Place the coop in a zone that naturally limits dog access.

Helper and emergency access: You will eventually go out of town, get sick, or face a household emergency. When this happens, someone else will need to manage your flock. A pet sitter, a helpful neighbor, or an emergency responder must be able to navigate your setup easily. Can a helper find the feed bin, operate the gate latches, and collect the eggs without an hour of specialized training? Can they see what they are doing safely? A coop hidden behind a labyrinth of garden gates or requiring complex physical maneuvering is a liability. The setup must be straightforward, accessible, and intuitive.

Make the final placement decision

It is time to choose. There is no such thing as a perfect site. Every location in a suburban yard requires a tradeoff between space, convenience, and visibility. Your goal is to choose the best available site and resolve minor logistical issues before you build.

To finalize your decision, use the placement scorecard. First, rule out any unworkable sites using this automatic disqualification list. Cross a location off immediately if it:

- Violates local setbacks or utility-locate boundaries identified in Chapter 5.
- Sits in a low-lying area that collects standing water.
- Fails the 360-degree clearance test for repairs and clean-out access.
- Cannot be reached safely and easily in the dark.
- Places the odor path directly into a neighbor's open windows or primary patio.

Once you remove the automatic disqualifications, rate your remaining options.

Evaluate your remaining sites (Score 1 to 5 for each):

1. **Wheelbarrow and clean-out access:** Is there ample space to park a wheelbarrow and swing a shovel on the clean-out side?
2. **Feed and water workflow:** Is the carry distance for winter water buckets and heavy feed scoops short and straightforward?
3. **Nighttime visibility:** Is the walking path clear of trip hazards and easily illuminated by a motion light?
4. **Repair clearance:** Can you physically fit your body and tools around all four sides of the structure?
5. **Neighbor consideration:** Is the site positioned out of the prevailing wind path toward neighbor patios, and can it be visually screened?
6. **Helper and emergency access:** Could a pet sitter safely and easily operate the gates, locks, and nest boxes?

Tally the scores. The site with the highest score is your winner. It might not be the most aesthetically hidden spot in your yard, but it is the one most likely to keep your daily chores fast and your neighborhood peaceful.

If your winning location has a low score in one manageable category, perhaps it provides excellent neighborly distance but requires a slightly longer walk from the feed bin, note that specific issue now. You can usually solve a minor workflow problem by buying a secondary watertight feed can to keep closer to the run, or by installing better path lighting.

By placing the coop where you can reach it safely, maintain it properly, and manage the workflow efficiently, you are building the foundation for a successful flock. With your exact, practical footprint decided, you are ready to evaluate the physical coop structure itself.

Chapter 7: Choose a coop you can clean, secure, and maintain

You are browsing online or walking the aisles of a farm store when you spot it. It looks like a miniature storybook cottage, complete with crisp trim, a decorative roofline, and perhaps a tiny window box for flowers. The product description promises it comfortably holds up to six hens. It is charming, and you can picture it sitting perfectly at the edge of your garden.

But three months from now, on a rainy Tuesday morning, you will not care about the trim. You will be kneeling in the mud, trying to squeeze your shoulders through a tiny door, scraping wet bedding out of a dark corner you can barely reach. You will realize that the beautiful storybook coop was designed to look fantastic in a catalog, not to function in a real backyard. Pretty does not clean well.

One of the most common mistakes suburban beginners make is treating the coop like garden furniture rather than an animal housing system. A reliable backyard coop is not the cutest coop. It is the one you can clean, secure, inspect, and live with every single week. Evaluating a coop before you buy or build requires looking past marketing photos and examining how the structure will perform under the daily stress of weather, predators, and the messy reality of keeping livestock.

If a coop setup makes chores miserable, your flock's health and your property's sanitation will eventually suffer.

Capacity claims and the space illusion

The most misleading aspect of coop shopping is the stated capacity. You might see a small, telephone-booth-sized structure advertised to "Hold up to six hens!" These claims often reflect the bare minimum square footage a bird needs just to stand inside the box. They rarely account for space taken up by feeders, internal nest boxes, or waterers.

Crucially, they do not account for daily management or bad weather.

You will often hear universal "rules of thumb" suggesting 3 to 4 square feet of interior coop space and 8 to 10 square feet of outdoor run space per bird. While university Extension programs frequently use these numbers as a starting baseline, they require massive caveats based on your specific situation.

Space needs change depending on breed size, climate, and confinement time. Smaller bantam breeds require less room, while heavy breeds like Orpingtons require more. If you live in a snowy climate where your chickens may refuse to leave the coop for days at a time, providing only the minimum interior space will quickly lead to cramped conditions. When hens are crowded, stress behaviors like feather-pecking and bullying can erupt rapidly. Furthermore, local municipal codes or homeowner's association (HOA) rules may mandate specific square-footage minimums per bird, regardless of what a manufacturer claims.

Always view marketing capacity with skepticism. Evaluate the space based on your planned flock size, the maximum amount of time birds will spend locked inside during bad weather, and your local ordinances.

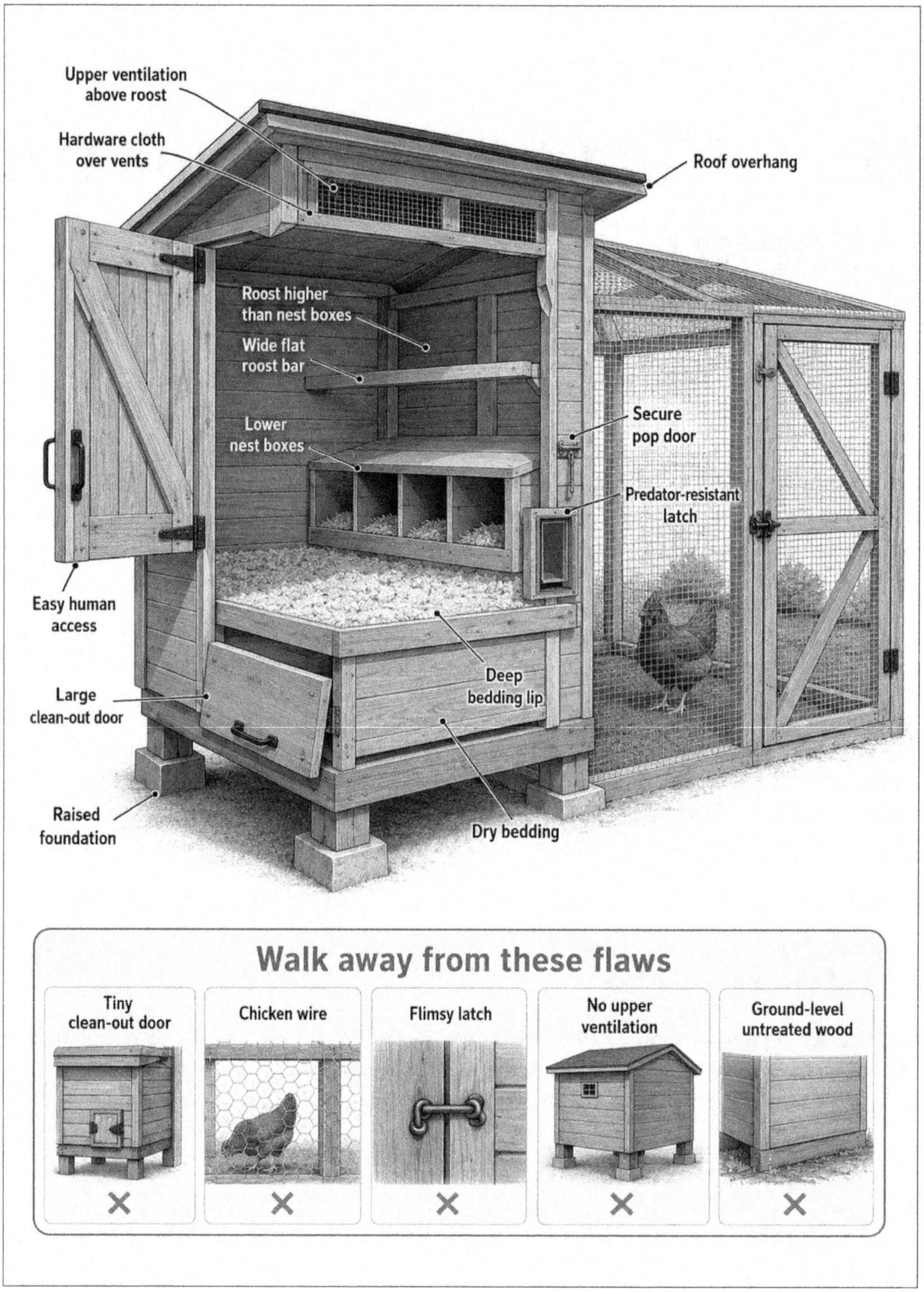

Functional Coop Cutaway

Cleanability and the bathroom analogy

Think of a coop like a bathroom you are personally responsible for cleaning every week. If that bathroom had a door only two feet high, and you had to crawl to wipe down a floor you could barely see, you would quickly resent it. Human access is critical. If you cannot reach it, you cannot maintain it.

Many prefabricated mail-order coops are scaled down entirely to the size of the chickens. This makes flat-pack shipping cheaper for the manufacturer, but it ignores the person doing the chores. Use this mental prompt before committing to a design: Show me where your hand, scraper, bedding bag, and flashlight will go.

Picture the exact cleaning route. You want large clean-out doors that swing wide open, allowing you to sweep soiled bedding directly into a wheelbarrow without taking the structure apart. You must also consider maintenance clearance outside the coop. If you place a coop two feet from a fence, you will not have enough room to park a wheelbarrow beneath the clean-out door.

Pay close attention to bedding management. To manage moisture and waste, you will likely use layers of pine shavings, hemp, or chopped straw. A good coop design includes a deep lip, a board at least four to six inches high across the bottom of the doorways, to hold the bedding inside when you open the door. If a coop door is flush with the floor, expensive bedding will spill into your yard every time you check on the birds.

Also, be wary of the marketing claim that slide-out cleaning trays solve all maintenance problems. Shallow metal or plastic trays supposedly pull out for effortless dumping. In reality, droppings and bedding get wet, heavy, and compressed. Material spills over the edges, jamming the sliding tracks. Trays freeze stuck in the winter and warp in the summer. You usually end up scraping out the inside anyway, working around a broken tray.

Ventilation, ammonia, and frostbite

It is entirely natural to assume a coop needs to be an airtight, snug box to keep chickens warm. But a tightly sealed coop traps danger inside.

Chickens naturally run hot, and they constantly add moisture to the air through their breathing and their wet droppings. If that damp, stale air cannot escape, it builds up. When cold temperatures hit, this trapped humidity is the primary cause of frostbite on a chicken's comb and wattles. Frostbite is rarely caused by cold air alone; it is caused by freezing moisture settling on the bird in a poorly ventilated space.

In warmer weather, a sealed coop traps heat and ammonia gas emitted by decomposing manure. High ammonia levels will quickly burn your hens' respiratory tracts and eyes, leaving them highly vulnerable to respiratory diseases.

To prevent these issues, coops require constant upper-level ventilation. This usually means generous openings near the highest point of the roofline (like soffit or ridge vents), well above the chickens' heads when they are sleeping. The goal is to let warm, stale, moisture-laden air rise and exit, drawing in fresh air without blowing a chilling breeze directly across the roosting birds.

These ventilation gaps cannot be left open to predators; they must be permanently secured with heavy-duty hardware cloth. Because conditions vary wildly by geography, check with your local agricultural Extension office for specific regional guidance on balancing winter ventilation with cold-weather protection.

Roosts, nest boxes, and interior layout

The interior does not need to be complicated, but it must respect chicken biology and behavior.

Chickens instinctively seek the highest available point to sleep to avoid nighttime predators. Because of this, their roosting bars must be positioned higher than their nest boxes. If a coop is designed so the nest boxes sit higher than the roosts, your hens will likely sleep inside the nests. Because chickens produce a significant amount of manure while sleeping, they will heavily soil the exact space where you want them to lay clean eggs, turning daily egg collection into a filthy, unsanitary chore.

Check that the roosting bars are sturdy. A common Extension-backed recommendation is to use standard 2x4 lumber laid flat, with the top edges slightly sanded or rounded off. Unlike wild songbirds that wrap their toes around thin branches, heavy laying hens rest flat on their feet. A wider surface allows them to sit fully down, covering their toes with their breast feathers to protect against frostbite in winter. Plan for roughly 10 to 12 inches of roost space per bird, and ensure the roost is spaced far enough from the back wall so the birds do not rub their tail feathers against the wood.

Nest boxes should be positioned lower, in a darker, quieter area of the coop. You do not need a separate nest box for every hen. A standard ratio is one box for every three to four hens. Even if you provide six identical boxes, you will frequently find three hens crammed into a single favorite box while the others sit empty.

Roof design and weather protection

A coop roof must confidently shed water away from the structure. Flat roofs or roofs without adequate overhangs will eventually allow rain to seep down the exterior walls, rotting the wood and soaking the bedding inside. Wet bedding breeds bacteria, flies, and odor.

Look for roofs that extend several inches past the coop walls, effectively protecting the ventilation gaps, pop doors, and windows from driving rain. The roofing material itself matters. Asphalt shingles, corrugated metal, or heavy polycarbonate panels are reliable. Thin wood or decorative roofing felt will warp, leak, and require replacement within a year or two.

Predator resistance and hardware

Suburban environments are full of patient predators. Raccoons, foxes, neighborhood dogs, and clever rodents have plenty of hours in the dark to test boundaries while you sleep.

There is no such thing as a completely "predator-proof" coop. Given enough time and desperation, wildlife can defeat almost anything. Your goal is predator “resistance”, building a structure tough enough to frustrate predators until they give up and move on.

Appendix F: Predator Resistance Audit has more info.

Inspect the closures on doors, windows, clean-out hatches, and nest box lids. A latch is not decorative; it is a critical component of your safety system. Surprisingly, many expensive commercial coops use lightweight hardware. A simple wooden toggle (turn-button) that twists to hold a door shut is completely inadequate. Raccoons have highly dexterous front paws and can easily twist these open. Basic hook-and-eye latches can often be jostled or bumped loose.

Start strong. Use heavy-duty slide bolts, carabiner-secured latches, or closures that require a complex, two-step motion to open. Check the hinges to ensure they cannot be easily pried off the wood.

The reality of prefabricated coops

As you evaluate designs, you will likely weigh buying a prefabricated mail-order kit against building from scratch or converting a sturdy garden shed.

Buying a prefab coop is faster, but an expensive price tag does not guarantee quality. Many flat-packed kits are constructed from thin fir or softwoods that splinter, warp, and split after a single wet season. Their wire panels are often fastened with weak staples that a large dog or raccoon can push straight through.

Sometimes, local farm stores sell excellent shed-style wooden coops built by nearby carpenters, which often dramatically outperform flat-packed kits. Alternatively, modifying a standard, human-sized garden shed into a coop ensures you have proper framing, heavy doors, and a roof you can stand under while cleaning.

Building or converting allows you to customize the footprint to fit an awkward corner of your yard, but it requires an honest assessment of your carpentry skills, available time, and tool access. Whichever route you choose, verify whether the final structure’s footprint triggers municipal building permits, setback requirements, or HOA architectural reviews.

Walk-away triggers

Whether staring at a catalog, standing in a farm store parking lot, or browsing online listings, you need a ruthless evaluation method. Walk away from any coop that features these critical flaws:

1. **You cannot reach the back corners.** If you have to crawl inside to retrieve a sick bird or scrape out a dirty corner, do not buy it.

2. **Chicken wire instead of hardware cloth.** Chicken wire keeps chickens in, but it does not keep predators out. Dogs and raccoons can tear through it. Any ventilation or window opening must be secured with half-inch, welded hardware cloth.

3. **Flimsy latches.** If it secures with a simple wooden spinner or a weak hook, it is a risk.

4. **No upper ventilation.** If it is a completely sealed box, it will trap ammonia and moisture.

5. **Ground-level untreated wood.** Wood that sits directly on wet soil will rot quickly and invite tunneling rodents. Coops need to be elevated on blocks, legs, or a dedicated foundation.

6. **Nest boxes higher than roosts.** You will be cleaning manure off your eggs daily.

Use this prompt when evaluating any design: Identify every opening a predator, rodent, driving rain, or spilled bedding could exploit.

Choosing the right housing is not about finding the prettiest yard ornament. It is about selecting a structure that makes keeping a flock a manageable, sanitary routine rather than a daily struggle against mud, disease, and poor design. Once you have a cleanable, resistant coop selected, the next step is making sure your hens have a safe place to stretch their legs. The nighttime shelter needs an outer perimeter, which means it is time to build a run that safely contains your birds and keeps the neighborhood out.

Chapter 8: Create a run that contains hens and resists predators

Understand this from the start: no setup is entirely predator-proof. Given enough time, hunger, and opportunity, nature will test any barrier you build. Your goal is not an impossible guarantee. Your goal is to build a highly predator-resistant system that forces wildlife to give up and look for an easier meal elsewhere.

Many beginners assume that a basic wooden frame wrapped in thin wire is enough to keep a flock safe. But a run is not just a fence designed to keep your birds from wandering into the street. It must function as a defensive structure.

When you stop viewing the run as a simple boundary and start treating it as a predator-resistance system, you replace vague anxieties with a practical habit of inspection and prevention.

Containment versus resistance

There is a distinct difference between keeping chickens in and keeping predators out. Containment requires very little effort. A standard wooden privacy fence or a flimsy roll of garden wire might successfully keep a hen from scratching in your neighbor's mulch. But containment materials fail under pressure.

Hens need contained outdoor access to scratch, dust bathe, and experience fresh air. While letting them wander freely across your property, a practice known as free-ranging, is highly popular on social media, it is rarely realistic in a standard suburban setting. Free-ranging provides zero predator resistance. Supervision breaks down when a phone rings, a child needs attention, or you step inside for five minutes. It only takes seconds for a neighborhood dog to jump a low fence or a hawk to strike.

A spacious, fully enclosed run protects your yard, protects your neighbors, and protects your birds. It ensures that when you leave for work or go inside to make dinner, your flock remains exactly where they belong, protected by tested barriers rather than luck.

Common suburban predators

Many people look around their manicured subdivision, see sidewalks and streetlights, and think, "We do not have predators here." But wild and domestic animals routinely forage in residential neighborhoods, particularly at night. To build a resistant run, you must understand the categories of risk your setup will face.

First, digging animals like foxes, skunks, coyotes, or loose dogs often attempt to tunnel under a fence line. Second, climbing animals like raccoons, opossums, and domestic cats scale wire walls to look for overhead openings. Third, reaching animals, specifically raccoons, will shove their dexterous paws through wide wire gaps to grab whatever is sleeping near the edge. Fourth, flying predators, such as hawks and owls, strike from above. Finally, squeezing predators like snakes, rats, and weasels are drawn to eggs, spilled feed, and vulnerable chicks, slipping through surprisingly small gaps.

Always research your specific regional caveats. If you live near wooded areas with bears, fishers, or heavy coyote populations, standard suburban setups may not be enough, and you may need to consult local Extension offices regarding heavy-duty electric fencing.

Chicken wire versus hardware cloth

One of the most dangerous misconceptions in backyard poultry is the belief that chicken wire keeps chickens safe. The logic seems sound: it is called chicken wire, so it must be for chickens.

Chicken wire is traditional hexagonal poultry netting made of very thin metal. It is strictly a containment tool. A determined raccoon, dog, or fox can push, stretch, or chew right through it.

To create actual predator resistance, Extension and poultry housing experts consistently recommend heavy-duty hardware cloth. Hardware cloth is a welded wire mesh. For primary run walls and ventilation coverings, a common recommendation is half-inch, 19-gauge hardware cloth, but verify local Extension guidance and your regional predator risks before relying on any single specification.

The half-inch mesh helps stop a raccoon from reaching its paws through the barrier, and it deters many snakes and rodents. Because the wires are 19-gauge thick and welded together at every single intersection, they do not pull apart when a dog jumps against them or a raccoon hangs from them.

However, strong wire is only as effective as its fasteners. If premium hardware cloth is secured with standard office staples, loose screws, or attached to a flimsy, rotting frame, the barrier will fail. When predators pull on the wire, standard staples easily rip out of the wood. You must verify that your wire is attached using heavy-duty pneumatic construction staples or thick exterior screws paired with wide fender washers, driven deeply into sturdy framing lumber.

Digging defense and utility safety

Because digging predators will attempt to tunnel under the run, the point where your wire meets the dirt is highly vulnerable. Simply resting a wooden frame on the grass offers no resistance.

There are two primary ways to deter digging. The traditional method is to dig a trench around the perimeter and bury the hardware cloth straight down, typically 12 to 18 inches deep. While highly effective, trenching is labor-intensive, especially in rocky or root-filled suburban soil.

Utility locate warning: Before digging post holes, trenches, predator aprons, or fence lines for a chicken run, contact 811 in the U.S. or your state 811 website/local one-call center and wait the required time for utility responses, markings, or clearance. Outside the U.S., use the local utility-locate service. Striking buried gas or electric lines can cause serious injury or death; damaging any underground utility, including water or sewer lines, can cause outages, property or environmental damage, and costly repairs, fines, or liability. Check state and local rules, and arrange private locating for private lines that 811 may not mark.

The second, increasingly popular method is an anti-dig apron, sometimes called a predator skirt. Instead of digging a deep trench, a wide strip of hardware cloth, usually 18 to 24 inches wide, is securely fastened to the bottom edge of the run. This wire is laid flat on the ground, extending outward away from the coop. You pin it down tightly with heavy landscape staples and cover it with a thin layer of topsoil or mulch.

Grass will quickly grow through the mesh, locking it in place. When a predator approaches the run, it typically walks right up to the vertical wall and starts digging at the base. Because their paws hit the buried wire apron, they cannot dig down, and they rarely think to back up two feet to start their tunnel.

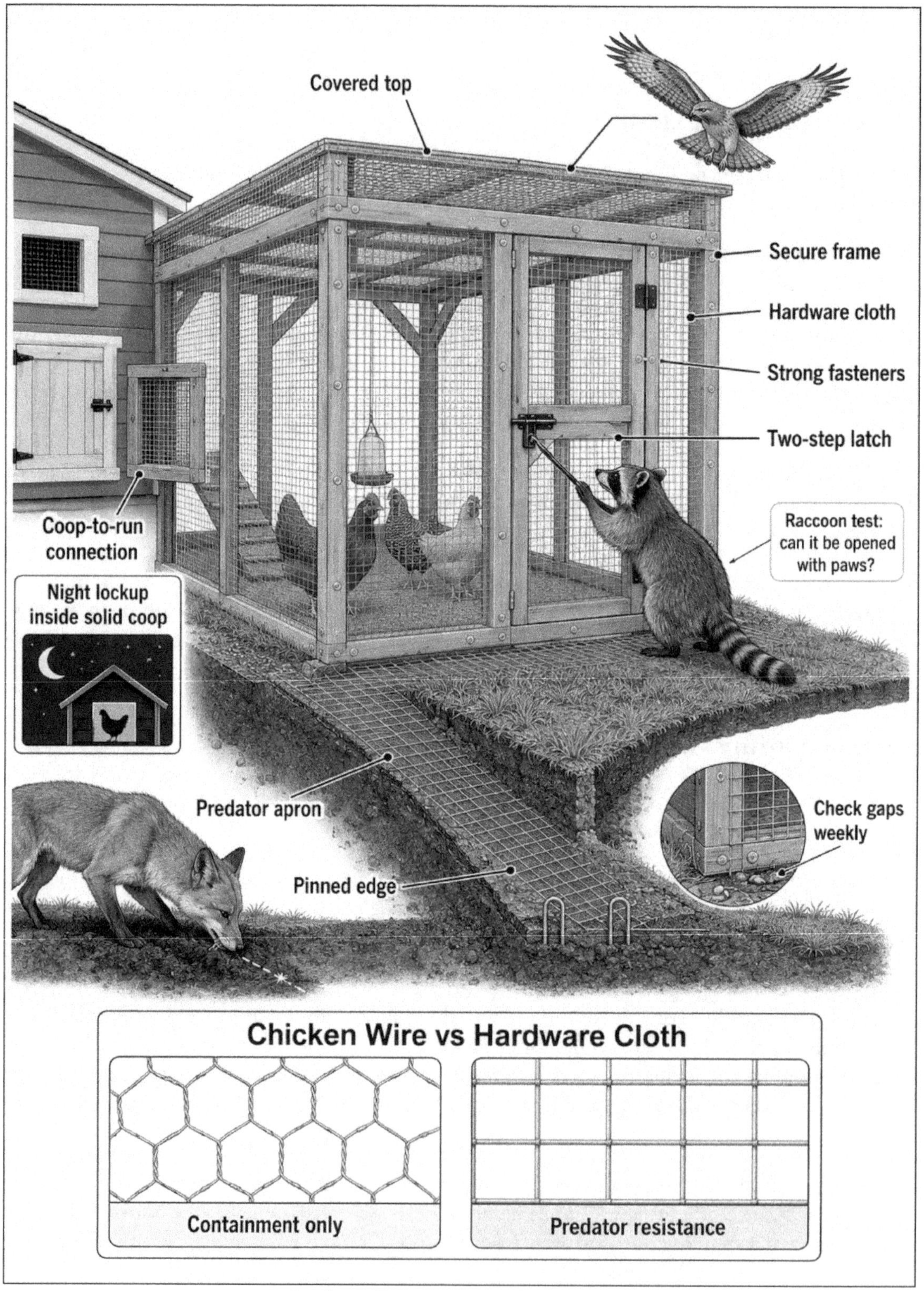

Predator-Resistant Run Cross-Section

Top coverage and overhead risks

A resistant run needs a secure top. Leaving a run uncovered invites several risks: your more athletic hens may fly out, climbing mammals will drop in, and hawks are given an unobstructed flight path.

You have three main options for top coverage.

Aviary netting: Heavy-duty, properly installed aviary netting is an affordable way to stop hawks and keep chickens contained. However, netting is not predator-resistant against climbers. A raccoon can easily chew or tear its way through a netted roof, and netting provides zero protection from rain or snow.

Hardware cloth: A wire top made of welded hardware cloth secures the run against climbing predators, provided the wooden frame can support the material. Like netting, it offers no weather protection, leaving the run exposed to mud and moisture.

Solid roofing: A solid roof, such as corrugated metal or polycarbonate panels, keeps the run drier, provides essential shade, and blocks aerial and climbing predators. Drier runs smell better and are much easier to manage. However, a solid roof catches the wind and accumulates heavy snow. It requires proper slope, structural framing, and anchoring to ensure it does not collapse or blow away.

Choose top coverage that addresses your local weather and specific neighborhood predator risks, while strictly following the structural limits of your coop's design.

Portable electric netting caveats

Many suburban beginners are targeted by advertisements for portable electric poultry netting. These systems use a highly visible, electrified mesh fence attached to step-in posts, powered by a solar or plug-in energizer.

Electric netting can be an excellent tool for daytime containment and deterring ground-based predators like dogs or coyotes, but it comes with strict limitations. First, it is not a roof; it offers no overhead protection against hawks or owls. Second, climbing predators can bypass it by dropping in from overhanging tree branches. Third, the netting must be kept clear of tall grass and weeds; if wet vegetation touches the lower wires, the fence shorts out and loses its charge.

Finally, electric fencing may be banned or restricted by suburban municipalities or homeowner association guidelines due to liability and safety concerns regarding neighborhood children and pets. Do not treat electric netting as your sole, overnight predator-resistance strategy in a suburban environment without checking local rules, site conditions, and manufacturer safety guidance.

Latches and the raccoon test

Even the strongest hardware cloth and the widest wire apron offer little protection if the run door is secured with a flimsy latch. When choosing hardware, keep the "raccoon test" in mind. Raccoons possess highly dexterous front paws and are persistent, intelligent problem-solvers. They can manipulate simple objects, easily flip open a standard hook-and-eye latch, turn a simple wooden block, or lift the lid off an untethered feed bin. To resist this kind of persistence, aim for secure, two-step closures. Carabiners attached to heavy-duty sliding bolts, spring-loaded locking latches, or padlocks are the safest options.

However, there is a catch: the latch must be easy enough for you to use every single day. If your locking mechanism is so complex or stiff that you dread using it, human nature will eventually take over. You might tell yourself, "I will close them in most nights, but tonight I am too tired to fiddle with that padlock in the rain." A predator only needs one forgotten night. Make your lockup routine deliberate but frictionless. Consider guest access as well; if a neighbor is checking on your flock while you are on vacation, the latch must be intuitive enough for them to secure properly.

The dog dynamic

Before reviewing your physical hardware, pay special attention to one specific suburban risk category: dogs. In many residential environments, a major threat to a flock is not wildlife. It is a loose neighborhood dog, or worse, your own family pet. Consider a common scenario: a family brings home a flock and assumes their gentle, older retriever will welcome the birds. The dog has never shown aggression toward humans. But chickens flutter, squawk, and run erratically. This movement triggers a deeply ingrained prey drive. The dog does not necessarily want to attack; it wants to catch the fluttering toy. In a matter of seconds, play turns fatal.

Do not assume a friendly dog will automatically be safe with chickens, and do not let obedience training claims replace physical containment. "My dog is friendly" is a wonderful truth for your household, but it is not a predator-resistance strategy. Your run should be strong enough to withstand a heavy dog jumping against the wire. If your own dog, or a dog on the other side of a shared property line, obsessively paces and barks at the run, you should add a visual barrier, like a solid lower wooden wall or a shade cloth, to break their line of sight. Removing the visual trigger significantly reduces stress for both the dog and your hens.

Finding the predator gaps

Predator resistance is not a project you finish once; it is a standard you must maintain. Over time, ground settles, wood warps, and hinges sag. You need to know exactly where the weak points are.

Any gap larger than half an inch is a potential failure point. If you can slide your thumb into a gap, a snake, a rat, or a weasel can get through it. If a raccoon can reach its fingers through a gap, it will pull at whatever is on the other side.

As you inspect your setup, look closely at these common points of failure:

- **Corners:** Check where two panels of wire meet. If the wood shrinks, a gap opens.
- **Doors:** If your door sags on its hinges, the space between the door and the frame might widen.
- **The base:** Check where the run meets the soil, watching for erosion, washed-out dirt, or fresh digging attempts.
- **The coop connection:** Inspect the area where the run attaches to the coop itself. Beginners often build a decent run and then leave a massive gap where the two structures meet.
- **The roofline:** Check where wire meets the rafters. Corrugated roofing often leaves wide, open channels at the eaves where rodents can slip inside.

The nighttime routine and inspection audit

The most dangerous time for your flock is between dusk and dawn. Relying solely on your run for nighttime protection is a gamble. Your daily operating plan should include a reliable nighttime lockup routine. Every evening, as the sun goes down, your hens will naturally return to the enclosed coop to roost. Once they are inside, make it your routine to go out and securely latch the solid coop door, shutting them safely inside the wooden structure, separate from the outdoor run. Even the best runs can be breached by a highly determined predator or a falling tree branch. Locking the hens inside the solid coop adds a critical second layer of defense.

Before you bring a single bird home, evaluate your yard with a critical eye using the tools in Appendix D: Coop/run inspection and predator audit. Change your perspective. Inspect the run as if you are looking for the easiest way to break in. Push firmly on the wire panels. Do they bow? Do the fasteners pop? Shake the coop connection to ensure the run is securely anchored. Get into a simple rhythm of inspection: check the perimeter before your birds arrive, after heavy rainstorms that might wash away soil, after you spot animal tracks, and as part of a weekly walkaround.

A highly resistant run gives your flock a secure space and helps you sleep at night. By expecting wildlife, planning for neighborhood dogs, choosing hardware cloth, securing the base and top, and building reliable daily lockup habits, you remove the anxiety of the unknown. But keeping trouble out is only half of the management process. Now that your birds are secure in their space, the next job is managing what accumulates inside that space.

To keep your neighbors happy and your yard safe, you will need to establish a system for handling feed, moisture, odor, rodents, and waste.

Chapter 9: Prevent odor, flies, rodents, and waste from day one

If you ask suburban homeowners what they fear most about a neighbor getting backyard chickens, they usually mention barnyard smells, swarms of flies, and rats. These are understandable concerns. However, chickens themselves are rarely the root cause of these problems. Management creates or prevents a neighborhood nuisance.

A common misconception is that a small flock will not smell simply because it is small. Three hens in a wet, unmanaged run can create strong odors and attract pests within days. Cleanliness must be a proactive routine, not an exhausting deep-clean you perform only when conditions become unbearable.

The most effective way to manage these risks is through Integrated Pest Management (IPM). IPM is a structured, environmentally sensitive approach that focuses on preventing pests by removing their access to food, water, and shelter, rather than immediately reaching for chemical sprays or poisons. By establishing IPM habits early, you turn what could be a stressful chore into a highly predictable system. Odor, flies, rodents, and waste are much easier to prevent from day one than to fix after a neighbor complains.

Moisture, stormwater, and odor

Dry, well-managed bedding is far less likely to smell than wet manure or soaked wood chips. Moisture is the engine that drives odor. Your primary job is to keep the coop and the outdoor run from becoming persistently wet.

Consider a common scenario: a beginner sets up a beautiful coop. After the first heavy spring rain, the yard smells terrible. The keeper assumes they need to clean more often, shoveling out wet dirt and adding fresh shavings every few days. But the real problem is not the cleaning schedule. The coop roof lacks a gutter, and runoff pours directly into the center of the run, soaking the bedding. Adding a simple gutter and downspout directed away from the run solves the moisture problem, and the odor disappears.

You must also consider where that runoff goes. Local stormwater, public-health, or environmental rules may restrict runoff that leaves your property or enters storm drains. Runoff that flows through a chicken run can pick up excess nitrogen, phosphorus, and bacteria, so design the site to keep dirty water out of storm drains, neighboring properties, and clean waterways. To prevent odor and stormwater contamination, place your coop and run where airflow and drainage naturally work in your favor, and use landscaping or grading to direct clean rainwater away from the poultry area entirely.

Inside the coop, watch the corners, the areas directly beneath the roosting bars, and the space around the waterer. A waterer that leaks every time a hen drinks creates a localized wet spot that quickly begins to ferment and smell. You cannot guarantee a perfectly odor-free keeping experience, but keeping the environment dry usually prevents odors from reaching the fence line.

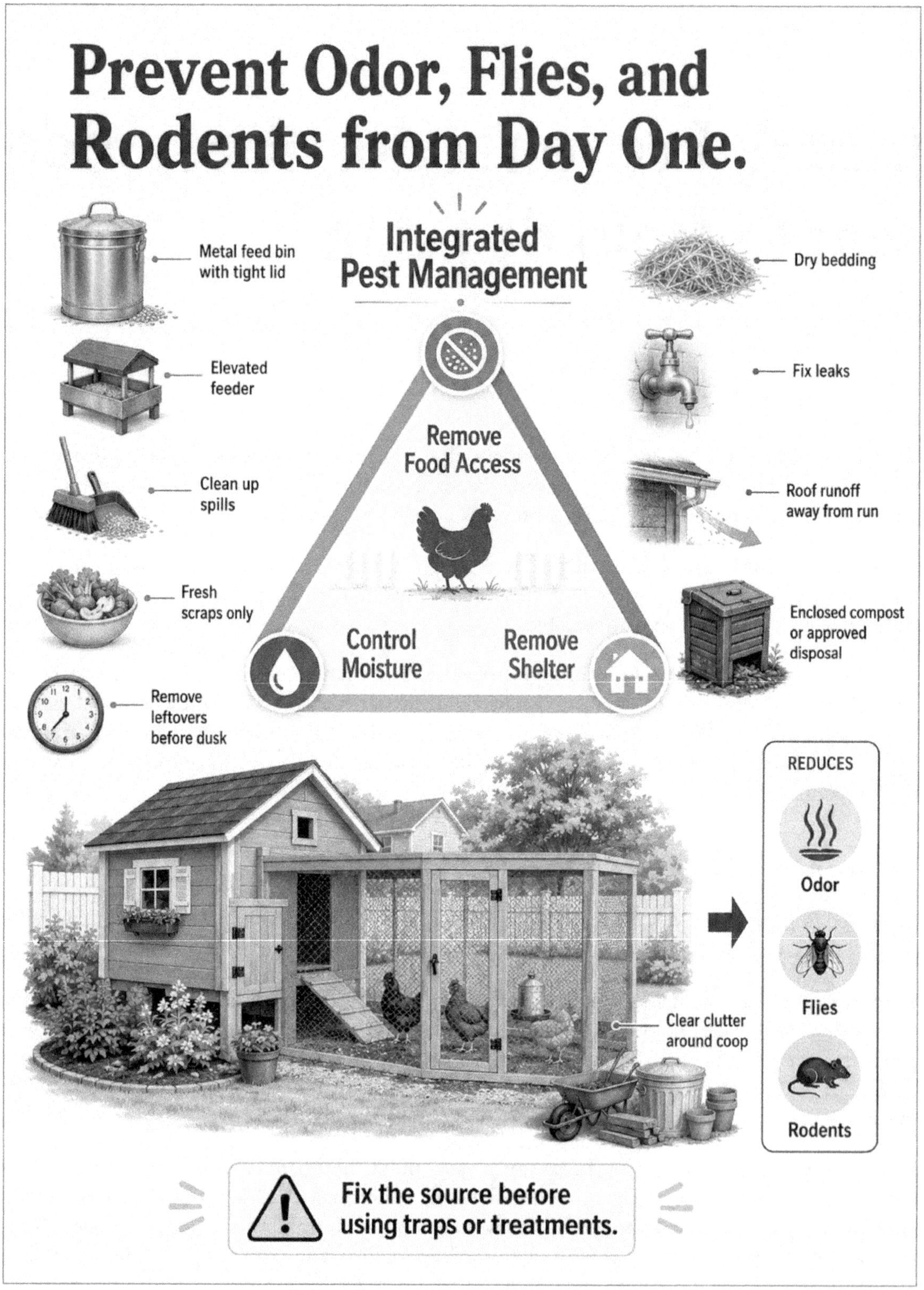

Odor, Flies, and Rodents IPM System

Flies and safe scrap management

Flies are opportunistic. They seek out wet, rotting organic material to lay their eggs in, alongside easily accessible food. Under IPM principles, your first line of defense against flies is sanitation, removing the environments where they reproduce.

Rake out deeply wet bedding immediately, especially during the warm summer months. Do not let spilled feed sit on the ground and turn into a paste after a rainstorm.

Kitchen and garden scraps require strict management. Beginners often enjoy treating their flock by tossing vegetable peels or leftover fruits into the run. However, leaving scraps to sit in the summer sun for days attracts flies rapidly. Furthermore, moldy, spoiled, or rotting scraps can be a direct threat to your flock.

Do not feed chickens moldy, spoiled, or rotting scraps. Moldy feed or food may contain mycotoxins that can make poultry sick or, in severe cases, contribute to death; spoiled or decaying organic material, contaminated feed, carcasses, or maggots can also expose birds to botulinum toxin, the cause of avian botulism. When in doubt, throw it out and provide fresh feed. If you provide fresh, safe scraps, only offer what the hens can finish in ten minutes, and strictly remove any uneaten remnants before nightfall.

If you choose to use fly traps, place them far away from the coop. Traps are designed to attract flies, so hanging one directly above the run will draw more pests into the birds' living space. If you feel chemical fly-control products are necessary, choose only products whose label specifically allows the intended use around poultry housing, and follow the label. Avoid household insect sprays and agricultural chemicals unless the label specifically allows use in a backyard coop; off-label use can harm the birds, contaminate the eggs, or violate pesticide rules.

Rodent prevention and pest control

A persistent myth in backyard poultry is that keeping chickens automatically guarantees a rat or mouse infestation. This leads to an unfortunate emotional block: beginners often think seeing a rodent means they are failing as a chicken keeper.

Rodents are usually attracted less by the hens themselves than by spilled feed, shelter, and easy access. Take the case of a suburban keeper who noticed mice darting under the coop at dusk. Panicked, they assumed the neighborhood was infested. A closer look revealed their hanging feeder was slightly tilted, allowing layer crumbles to cascade onto the ground. The mice were living off the spill. Once the keeper switched to a spill-proof feeder design and swept up the yard, the rodent activity dropped significantly.

Preventing rodents requires disciplined feed control. Ask yourself right now: Where will feed be stored so children, guests, pets, and pests cannot access it? Leaving a paper sack of chicken feed in a partially open garage or an unsealed plastic tub is an invitation to local wildlife. Feed must be stored in secure, tight-fitting metal containers. Galvanized steel trash cans with tight lids are a reliable choice.

Beyond feed, rodents look for places to hide. Keep the area around the coop free of stacked debris, tall weeds, piled lumber, and clutter that creates protected nesting spaces.

A severe caution on traps and rodenticides

If rodents appear despite your best sanitation efforts, IPM protocols dictate moving to exclusion (sealing holes with hardware cloth) and mechanical control (traps) before chemical control (poisons).

Be highly critical of casual poison recommendations found on internet forums. Rodenticides pose severe risks to the surrounding environment. If a poisoned mouse staggers out into the yard, it can be eaten by a family dog, a neighborhood cat, or a predatory bird like an owl or hawk, potentially causing secondary poisoning. Chickens themselves will readily hunt and eat a slow-moving, poisoned mouse, which can sicken the hen or create egg-safety concerns.

If you use traps or rodenticides, use only methods and products appropriate for the specific site, follow product labels and manufacturer instructions exactly, and check whether state or local rules add restrictions. If rodenticides are used at all, use only products labeled for the specific site. Place rodenticide bait only in appropriate tamper-resistant, weather-resistant bait stations when required by the label or whenever children, pets, chickens, livestock, or wildlife could access the bait, and secure the bait inside the station.

No bait station should be treated as tamper-proof, and you should not assume any station absolutely prevents access.

Additionally, you need a plan for the safe, prompt removal and disposal of dead rodents in a manner allowed by the product label and local rules, to reduce secondary poisoning or disease risks. Before utilizing any toxic control methods, consult your local Extension office or hire a licensed pest-control professional who has specific experience with poultry safety and local wildlife regulations.

Waste, compost, and manure pathogens

A small flock of hens will generate a surprising volume of soiled bedding and manure over a year. If you wait until you are holding a shovel full of dirty pine shavings to figure out where it should go, you are already behind. You need a verified waste destination plan before the birds arrive.

Many beginners assume they can simply pile waste in a back corner to compost. This is a significant risk. Untreated poultry manure is "hot", meaning it is high in nitrogen and will burn garden plants if applied directly. More importantly, poultry manure can harbor pathogens like Salmonella and Campylobacter.

Do not assume an unmanaged backyard compost pile has killed poultry-manure pathogens. Reliable pathogen reduction requires a well-managed pile that is large enough to heat, monitored with a compost thermometer, aerated or turned as appropriate, and held at pathogen-reducing temperatures for the required time. If you cannot verify the temperature and management history, treat the material like raw or uncomposted manure around food gardens. Dust and runoff from an open, poorly managed manure pile can blow or wash into your vegetable garden, a neighbor's yard, or a child's play area, creating a genuine public health hazard.

Because of the odor, fly, and pathogen risks, some municipalities and homeowner associations may forbid open manure composting. Check your local public-health or solid-waste guidelines for residential livestock.

If you are legally permitted to compost, ensure your bin is fully enclosed, securely managed, and placed far away from neighbor conflict zones, shared fence lines, and stormwater drainage paths.

If composting is not an option where you live, you must verify exactly how to dispose of the waste according to your municipal trash rules. Do not assume you can fill your green yard-waste bin with soiled pine shavings; many municipal composting facilities reject animal waste. Disposal may involve using heavy-duty contractor bags to securely contain the soiled bedding for regular residential trash pickup. Verify these trash-disposal rules with your local solid-waste authority before the birds arrive so your weekly cleaning routine is legally compliant from day one.

The weekly cleanliness walk-through

To keep your proactive system running, establish a weekly inspection routine. This takes just a few minutes and replaces the need for massive deep cleans. Incorporate this checklist into your regular yard maintenance:

- **Smell test:** Check for odors directly inside the coop, from your own patio, and standing at the fence line exactly where your neighbors stand.
- **Wet bedding check:** Look closely at the ground beneath the roosting bars and the waterer. Scoop out any damp material before it ferments.
- **Waterer leak check:** Ensure valves are not dripping and the container is structurally sound.
- **Stormwater visual check:** Confirm that rainwater is draining away from the run and that no manure runoff is leaving your property.
- **Feed spill check:** Inspect the ground under the feeder. If there is wasted grain, adjust the feeder height or change the design.
- **Feed storage check:** Verify that your galvanized storage container is undamaged and the lid is sealed.

- **Scrap removal:** Ensure any garden or kitchen scraps given to the hens have been entirely removed before nightfall to prevent mold and flies.
- **Rodent signs:** Look for droppings, small tunnels near the run base, gnawing on wood or plastic, or disturbed bedding.
- **Bait/trap safety check:** If utilizing pest control, verify that all bait stations or traps remain securely locked, label-compliant, undamaged, and placed to prevent access by pets, children, and chickens.
- **Waste destination:** Confirm that soiled bedding has been moved to your planned, legally approved waste destination, whether securely bagged for trash or moved to an enclosed compost system.
- **Fence-line visual check:** Look at your setup from the neighbor's perspective. A tidy setup builds trust and makes it easier for neighbors to see that the flock is managed responsibly.

Responding early

Maintaining your flock becomes much easier when you follow these routines. When a problem does pop up, handle it early.

When an odor suddenly appears, inspect for moisture and waste first. You might find a hidden leak or a pile of manure tucked beneath a roosting bar. Should flies arrive in larger numbers, check your feed stations, rake out damp bedding, and confirm no unsafe scraps are rotting in the sun. When rodents show up, immediately secure your feed storage, clear the run of clutter, and eliminate their food source before considering traps or poisons.

By keeping your setup dry, sweeping up spills, and moving waste to its planned destination on a reliable schedule, you protect your flock's health and maintain neighborhood peace. With this cleanliness system firmly in place, you can move forward and plan for your flock with practical confidence.

Part 3: Pick the right first flock

You are standing in your backyard looking at a finished structure. The hardware cloth is tightly secured with heavy-duty fasteners, the latches drop perfectly into place with a satisfying metallic click, and the roof is fully sealed against the impending rain. Inside the coop, the feeder hangs at just the right height, the waterer is filled and tested for leaks, and the nesting boxes are freshly lined with a thick layer of pine shavings. You have done the necessary, often tedious groundwork to reach this exact moment. You have verified your local municipal ordinances, checked your homeowner association rules, mapped the property lines to ensure proper setbacks, built a sensible budget, and constructed an enclosure designed to resist local predators and seasonal weather. The hard, abstract planning phase is finally behind you.

The time has come to bring this structure to life. But before you open a colorful poultry catalog, visit a local farm supply store, and start filling a shopping cart with exotic breeds, you need to pause and establish a highly practical philosophy for your bird selection.

When a beginner first decides to keep a backyard flock, it is remarkably easy to assume that a chicken is just a chicken. The excitement of finally acquiring birds often triggers an immediate, almost subconscious pivot from practical planning to emotional shopping. Soon, you are scrolling past pages of fluffy Silkies, crested Polish hens with wild topknots, and specialty birds promising a rainbow egg basket filled with pastel blue, olive green, or deep chocolate-brown shells. In the spring, farm supply stores feature large metal bins of peeping chicks under warm red lights, making it incredibly tempting to point at a few cute birds, buy a cardboard box, and take them home on a whim. But picking a flock based purely on novelty, aesthetics, or impulse is one of the fastest ways to derail a suburban chicken-keeping experience before it even formally begins.

To understand why this phase requires discipline, you have to understand the fundamental difference between a rural farm setting and a suburban backyard. In an agricultural setting where chickens have acres of pasture to roam, breed selection is often simply a matter of personal preference and aesthetics. Unlimited space is an incredible shock absorber for challenging poultry behavior. If a farm bird is naturally flighty, nervous, or easily startled, it has plenty of space to run away from perceived threats. If a bird is exceptionally loud, vocal, or chatty throughout the day, the sound dissipates over the open fields long before it reaches a human ear. If a bird is a fierce, aggressive forager that tears up the ground in search of insects, the earth is vast enough to recover quickly without becoming a muddy wasteland. Furthermore, in a large, free-ranging farm flock, if two birds do not get along, the submissive bird can simply retreat and establish physical distance. The sprawling environment naturally diffuses tension, absorbs noise, and hides destruction.

In a typical suburban yard, however, every single one of those inherent traits is heavily magnified. Your flock will live in a tightly defined, enclosed footprint, likely bordered by a standard six-foot privacy fence and a neighbor whose kitchen window is only a few yards away. Suburbia acts as an intense magnifying glass on poultry behavior. A high-energy, flighty breed that constantly paces the fence line or launches itself over the enclosure into your neighbor's prized vegetable garden is no longer just a quirky, energetic bird, it is an immediate neighborhood nuisance and a potential liability. A loud, highly vocal hen whose persistent cackles echo off the vinyl siding of the houses next door creates instant tension with the people living around you.

Most importantly, suburban confinement creates a social pressure cooker for the birds themselves. When you place chickens in a perfectly secure but dimensionally limited backyard run, they cannot easily escape one another. A breed that is naturally high-strung, or an incompatible mix of tiny, fragile ornamental birds and massive, assertive standard-sized hens, will inevitably experience severe social friction. Without the physical option to walk away and de-escalate, that friction often turns into chronic bullying, feather pecking, resource guarding, and severe physiological stress.

Because of this distinct magnifying effect, you are no longer shopping for novelty or treating your flock like a living garden ornament. You are shopping for compatibility, calmness, and quiet. You need practical, robust birds that are fundamentally suited to a confined footprint and close proximity to humans.

This realization should actually come as a massive relief. It removes the immense pressure of having to choose from hundreds of overwhelming catalog options and replaces it with a simple, logical, criteria-based decision process. You have already built a defined box, legally, financially, and physically. Now, your only job is to fit a sensible, manageable flock into that box. When your setup is realistic and your constraints are clear, bird choice becomes significantly easier and entirely less emotional.

In the upcoming chapters, we will break down the specific, practical decisions you need to make to acquire your first flock. We will explore the logistics of the different life stages available to you, comparing the messy, time-consuming reality of raising fragile chicks indoors against the convenience of purchasing older, fully feathered birds. We will examine how to evaluate breeds based on necessary traits, emphasizing confinement tolerance, docile temperaments, and climate readiness, rather than just the color of their feathers.

We will also establish starter flock templates to help you choose a safe, manageable number of birds that respects both your local ordinances and the physical capacity of your new coop. Finally, we will confront the realities of suburban zoning rules and poultry genetics, focusing on why beginners must carefully navigate the acquisition process to avoid accidentally bringing home a rooster, and why every responsible suburban flock owner needs a backup plan.

Picking your first flock requires putting aside the impulse for the flashiest birds and focusing entirely on the reality of daily care. A good suburban first flock is boring in the best possible way. It is stable, quiet, reliably contained, and suited to the exact weather conditions of your region. By shifting your mindset from emotional shopping to practical compatibility, you ensure that when you finally open the door to that beautiful new coop, the birds that walk inside are exactly the ones you want to live with, and the ones your neighbors are happy to live next to, for years to come.

Chapter 10: Decide between chicks, pullets, and adult hens

IF YOU SPEND ANY time researching backyard chickens, you will eventually see the perfect photograph. A small child sits on a sunlit floor, gently cupping a fluffy yellow chick in their hands. It looks peaceful, simple, and wholesome. What the photo leaves out are the pine shavings kicked across the room, the fine layer of poultry dander settling on the baseboards, the heat lamp clamped to a plastic tub, and the loud chirping ringing through the house at two in the morning.

There is a stubborn misconception that real chicken keeping must start with chicks. This creates unnecessary pressure on beginners. You might assume that unless you raise your birds from day one, you are cutting corners. But in a residential neighborhood, the goal is not homesteading authenticity. The goal is a manageable, healthy flock.

To do that, you must choose the starting age that actually fits your household's time, space, and risk tolerance. You have three distinct entry pathways: chicks, pullets, and adult hens.

The three starting options

Before making a decision, you need to understand the distinct life stages of poultry.

Chicks are babies, typically ranging from day-old hatchlings to six or eight weeks of age. They are entirely dependent on you for artificial heat, specialized starter feed, and strict protection from drafts, dampness, and household pets.

Pullets are adolescent female chickens. While the technical definition applies to a hen under one year of age, backyard sellers usually use the term for a fully feathered bird sold between 12 and 20 weeks old. Depending on the weather, they are ready to live outside in a coop and are approaching egg-laying age (often called "point-of-lay").

Adult hens are mature females that are over a year old and have usually laid for at least a season. They are fully grown with established behaviors and immune histories.

Starting with chicks

Raising chicks is the most popular way to start, and they remain an excellent choice for prepared households. Farm stores and mail-order hatcheries offer a wide selection, allowing you to choose specific breeds you might not find locally as adults.

However, do not assume chicks are the ultimate budget option. While a day-old chick is inexpensive, raising them indoors requires a dedicated setup. You will need a secure enclosure, specific chick-sized feeders and waterers, starter feed, and a reliable heat source.

The challenges of the indoor brooder phase are immediate. As chicks grow their first real feathers, they shed heavy dander that covers nearby surfaces. You must also manage artificial heat safely. Heat lamps can start fires if they fall or sit too close to bedding or other combustibles. If using one, use a brooder-rated lamp with a porcelain socket and guard, hang it from a secure chain or cable loop rather than the cord or clamp alone, follow the lamp and bulb manufacturer's wattage and clearance instructions, keep it well away from bedding, cardboard, hay, and other combustibles, and position it where children, pets, or chicks cannot knock it down. Consider safer radiant heat plates or pads designed for brooding.

In a busy household, you must consider the dynamics of children and pets. All live poultry can carry Salmonella and Campylobacter. Strict hand-washing is mandatory every single time the birds or their equipment are touched. Toddlers are especially vulnerable because they touch their faces frequently and cannot follow hygiene rules reliably. If you have cats or prey-driven dogs, the brooder must feature a heavy, clamped, unbreakable lid.

Finally, starting with chicks involves sexing uncertainty. When you buy "sexed pullets," the hatchery has used vent or feather sexing to separate females from males. This process is typically about 90 percent accurate [VERIFY]. If you bring home a small flock of six chicks, there is a statistical possibility that one will grow up to crow. (Note: "Straight run" chicks are entirely unsexed, meaning you will likely get 50 percent roosters. Suburban keepers should never buy straight-run chicks.)

Starting with pullets

If an indoor brooder is not a good fit for your home, pullets are a highly practical alternative. Buying pullets is not cheating; it is a strategic decision to bypass the dust, heat-management, and indoor-hygiene risks of the baby phase.

A family with young children might choose pullets specifically to avoid keeping poultry in the house. Because pullets are fully feathered, they can usually move directly into a secure outdoor coop. Because their physical characteristics are developed, the risk of accidentally buying a rooster is significantly lower. They are also closer to laying age, meaning the wait for your first eggs may be a matter of weeks rather than months.

The main challenge with pullets is upfront cost and narrower availability. A quality pullet costs significantly more than a day-old chick, reflecting the weeks of feed, electricity, and labor the breeder has already invested. You must rely entirely on what local farms or breeders have chosen to raise.

Transporting adolescent birds requires sturdy, ventilated transport boxes or hard-sided dog crates. They will need a few days of quiet adjustment once they reach your yard. Most importantly, your permanent outdoor coop and run must be completely built, predator-resistant, and fully stocked before pullets arrive.

Starting with adult hens

Starting with adult hens usually happens through rehoming. A neighbor may be moving away, or a local farm may be downsizing its laying flock. Taking in healthy adult hens gives an existing flock a good home without the wait of raising young birds, and you will likely get eggs immediately.

Yet, people falsely assume adult hens are the easiest route. While physically hardy, adult birds come with history. You rarely know their exact age, which matters because a hen's egg production often drops significantly after her second or third year. They also arrive with established behaviors. If a hen has learned the habit of eating her own eggs or flying over a fence, she brings those habits to your yard.

Biosecurity is the greatest risk here. Moving older birds between properties carries the danger of transferring undetected respiratory illnesses, lice, or mites to your yard. Adult hens are best for highly cautious buyers who are comfortable walking away if a seller is evasive about the flock's health history.

The mixed-source danger and biosecurity rules

One of the most common, and devastating, mistakes a beginner can make is assembling a "rainbow flock" by buying two pullets from one farm, one hen from a neighbor, and three chicks from a store, then immediately putting them all together. Every farm has a different baseline of bacteria and viruses. Birds that are perfectly healthy on Farm A may carry a pathogen that Farm B's birds have no immunity against. When you mix sources immediately, you risk catastrophic flock loss.

Chicks vs. Pullets vs. Adult Hens Comparison

If you are starting your very first flock in an empty coop, buy all your birds from a single source on the same day. If you must add new birds to an existing flock, or if you are forced to buy from multiple sources, you must quarantine the new birds. Standard poultry biosecurity requires keeping new birds in a completely separate enclosure, ideally on a different part of the property, for at least 30 days. When moving between a quarantine pen and the main flock, wash or sanitize your hands and use separate, disposable, or cleaned/disinfected footwear. Ideally, dedicate separate boots, gloves, coveralls, and tools to each group.

Screening your source

Whether buying chicks or older birds, gather information before you hand over cash. Never assume the seller will guarantee health or proactively disclose disease history.

Questions for hatcheries and farm stores (Chicks):

- **Are these birds NPIP certified?** The National Poultry Improvement Plan (NPIP) certifies that the breeding flock was tested for specific diseases, primarily Pullorum-Typhoid, and sometimes Avian Influenza. It is an excellent baseline, but it does not guarantee the chick is free of all illnesses or respiratory infections.
- **Have they been vaccinated for Marek's disease?** Marek's is a devastating, highly contagious viral disease. For Marek's disease, plan ahead: vaccination is most effective when given in ovo or at hatch, before chicks are exposed to the virus. For most beginner backyard keepers, the safest and most practical option is to order chicks already vaccinated by the hatchery. If hatching chicks yourself, ask a poultry veterinarian or Extension specialist about proper vaccine handling, timing, and whether Marek's vaccination is appropriate for your flock.
- **What is your accuracy rate for sexed pullets, and what is your policy if a rooster slips through?**

Questions for local farms and breeders (Pullets and Hens):

- **What is the exact hatch date?** (Do not accept "about a year" if you want reliable layers.)
- **Were they vaccinated for Marek's at hatch? Can you provide records?**
- **Have these birds ever been treated for respiratory issues, mites, or lice?**
- **Has your flock ever experienced a sudden unexplained die-off?**
- **How will we transport them?** (Ensure birds from different farms are never placed in the same vehicle without physical and respiratory separation.)

Chicks vs. pullets vs. adult hens comparison

Factor	Chicks (0 to 8 weeks)	Pullets (12 to 20 weeks)	Adult Hens (1+ years)
Best for	Prepared households wanting specific breeds.	Cautious beginners and busy families skipping the brooder.	Rehoming scenarios, budget buyers.
Setup required	Secure indoor brooder, heat source, chick feeders.	Completed outdoor coop and predator-resistant run.	Completed outdoor coop and predator-resistant run.
Time to first egg	Longest wait (4 to 6 months).	Shorter wait (often weeks).	Immediate (if currently laying season).
Rooster risk	Highest (sexing mistakes happen at 10% rate).	Low (physical traits are visible).	Lowest (clearly identifiable).
Cost profile	Low per-bird cost, high equipment cost.	Highest upfront per-bird cost, no indoor equipment.	Usually cheap or free.
Disease/Health risk	High vulnerability to chilling; needs Marek's vax.	Moderate; depends heavily on breeder's biosecurity.	Highest risk of carrying hidden illnesses or parasites.

Completed busy-family decision example

To see how this decision plays out in reality, consider this example from a fictional household prioritizing safety and time. Use the blank *Entry-Pathway Decision Worksheet* in the companion PDF to map out your own scenario.

The Miller Household

- **Family logistics:** Two working adults, a three-year-old child, and a high-energy Labrador Retriever.
- **Housing status:** Local laws allow up to six hens. The outdoor coop and run are fully constructed and predator-resistant.
- **Brooder capability:** The only available indoor space is a guest bedroom.
- **Evaluation:** The Millers initially wanted chicks so their toddler could watch them grow. However, after reviewing the safety constraints, they realize an indoor brooder in a carpeted room creates unnecessary fire and dust risk. More importantly, keeping a three-year-old from touching the chicks (and risking Salmonella) while also keeping the Labrador out of the room would require constant policing.
- **The Decision:** The Millers choose **pullets**. They contact a local farm, reserve four 16-week-old vaccinated pullets, and bring them straight to the secure outdoor coop. They bypass the indoor hygiene risks entirely, avoid purchasing heat lamps, and eliminate the risk of a rooster waking the neighborhood.

The entry-pathway rule of thumb

Do not acquire any birds of any age until your local legal rules are verified and your infrastructure is ready. If your town requires an inspection before ownership, you cannot use an indoor brooder to bypass or delay this requirement.

If your household pets or young children create unmanaged risks inside the house, or if your weekday schedule leaves you no time to monitor a heat lamp, skip the chicks.

By removing the assumption that you must start with day-old hatchlings, you open up practical, safe ways to begin backyard chicken keeping. Whether you bring home a peeping box of fluffy chicks, a crate of awkward pullets, or a few seasoned adult hens from a verified source, the goal remains building a healthy flock.

Once you have decided on the right entry pathway, you are ready to tackle the next great temptation of the poultry world: deciding exactly which breeds will fill that coop.

Chapter 11: Choose calm, climate-sensible hens

You are sitting at your kitchen table late at night, a glowing tablet in your hands. You started out simply looking for "good backyard chickens." Now, two hours later, you are scrolling through dozens of online hatchery pages, staring at pictures of breeds you cannot even pronounce. There are birds with heavily feathered feet, birds with fluffy top hats, birds that lay olive-green eggs, and birds that look like miniature dinosaurs.

With every new tab you open, you become less certain. The options are paralyzing. You just wanted three friendly hens to put in the coop you mapped out. Now you feel like you need a degree in poultry genetics. Let's close the tabs and take a breath. Choosing your first flock should not feel like an overwhelming homework assignment.

Why this is not a breed encyclopedia

This chapter is deliberately not a breed encyclopedia. If you want a list of two hundred chicken breeds, complete with their historical origins and show-ring standards, there are massive, beautiful books dedicated to exactly that. But as a suburban beginner, treating a hatchery catalog like a checklist is often a trap.

The reality of a suburban backyard is grounded and practical. Your goal is simply a manageable, healthy, and low-stress first flock. You do not need every rare breed, nor do you need to preserve a heritage line imported from across the globe. What matters most is a bird's ability to thrive in the specific environment you are about to provide, and your ability to source that bird locally.

Choosing your first hens is much like choosing a first family dog. If you have a small, fenced yard and young children, a calm, predictable mixed-breed dog from the local shelter will almost certainly make your life happier than a high-strung, working-line purebred that requires three hours of daily running. Temperament and environmental fit always matter more than looks. You can narrow the field dramatically by focusing on a few practical, trait-based selection criteria.

Choose for your climate and frostbite risk

Your backyard weather dictates your birds' daily comfort. There is no universal chicken breed that thrives equally well in a blazing desert, a humid swamp, and a freezing mountain town. Many chickens tolerate cold better than they tolerate heat; they essentially wear down jackets year-round. However, high heat, especially combined with deep humidity, can be lethal to a flock. A chicken does not sweat. Instead, it uses its comb and wattles to regulate body heat, pumping blood into these bare patches of skin to radiate warmth away from its core.

If you live in a hot, southern climate, look for lighter-bodied breeds with larger, single combs. Breeds in the Mediterranean class, such as Leghorns or Andalusians, are highly heat tolerant. Heavy, dense birds with tiny combs will pant and struggle much faster through a sweltering July afternoon.

Conversely, if you live in a northern region with brutal, sub-zero winters, large single combs are highly prone to frostbite.

Frostbite is painful, damages tissue permanently, and opens the door to infection. In a cold climate, you want heavier, well-feathered birds with small, close-to-the-head combs, commonly called "pea," "cushion," or "rose" combs. Breeds like Wyandottes, Buckeyes, or Chanteclers sport these smaller combs and handle freezing temperatures with far less risk.

Keep in mind that breed choice does not replace the need for proper shelter. Heat, cold, and moisture management still depend heavily on you providing shade, ventilation, fresh water, dry bedding, and wind protection. Contact your local agricultural Extension office to find out what breeds reliably survive the hardest month of your local year.

Choose for confinement tolerance and temperament

After climate, confinement tolerance is the most important factor for suburban peace. Flighty, active breeds are wonderful survivors in a free-range farm setting where predators are a constant threat. But in a small suburban run, that high energy translates to stress. Active foragers will pace the fence line, quickly strip the yard of vegetation, and actively look for ways over the perimeter.

Heavy, traditional backyard breeds, such as Orpingtons, Australorps, or Sussex, are generally highly confinement tolerant. They are content to scratch in a defined space, taking life at a slower pace.

This slower pace usually translates to a calmer temperament. In a tight neighborhood, with children running next door, lawnmowers starting up, and the occasional dog barking at the fence, your chickens will experience a heavy amount of sensory input. Calm, docile breeds process this environment without panic. They are more likely to waddle over to see if you have a handful of feed, rather than scattering in terror every time you open the back door. This makes routine care, coop cleaning, and the occasional health check much easier.

Keep in mind that individual birds vary. Breed gives you a baseline tendency, not a guarantee. A naturally calm breed can become skittish if chased constantly by a toddler, just as a naturally nervous breed can become relatively tame with consistent, gentle handling. The daily environment you create shapes their behavior just as much as their genetics do.

Understand size and noise tendencies

Size: Standard-sized chickens need a specific amount of square footage. Some heavy breeds run exceptionally large, requiring more space on the roosting bar and wider nest boxes. Smaller birds, known as bantams, are roughly one-third the size of standard chickens. While bantams are easier to accommodate in a compact suburban coop and can be incredibly charming, they are highly vulnerable to neighborhood predators. A neighborhood cat or a small hawk that might ignore a heavy, eight-pound standard hen will actively hunt a two-pound bantam. For a beginner, standard-sized hens are generally much sturdier and easier to keep safe.

Noise: Hens are generally much quieter than roosters, but no chicken is completely silent. Some breeds are naturally more vocal, singing loud, repetitive "egg songs" after laying, or complaining loudly when they want a treat or are startled. Generally, the lighter, flightier breeds tend to be more vocal and easily alarmed, whereas heavy, calm breeds are more prone to quiet clucking.

Keep egg expectations realistic

Many beginners get fixated on finding the ultimate laying breed, hunting for birds that promise a massive annual yield. A hatchery catalog might promise "up to 300 eggs a year," but treating those marketing descriptions as a guaranteed contract will lead to disappointment. A beginner backyard is not a commercial, climate-controlled egg facility. A hen's laying frequency is dictated by her age, the season, the amount of daily sunlight, her nutritional intake, her stress levels, and her annual molt.

In her first year, a healthy hen of a standard laying breed will likely give you a steady supply of eggs. However, many hens slow down or stop laying entirely during their molt, dropping their old feathers to grow a new set for winter. During the darkest days of winter, laying will slow to a crawl or pause completely without artificial light. Even the most prolific breeds experience these natural, necessary breaks. Do not build your flock, or your household budget, around exact egg count promises. Encourage realistic expectations instead. If you choose calm, healthy, climate-appropriate hens, the eggs will come.

Avoid the novelty trap

As you narrow down your choices, beware of the novelty trap. It is tempting to order rare breeds with extreme traits simply because they look fantastic on social media. Birds with giant crests of feathers on their heads (like Polish chickens), fully feathered legs (like Cochins), or fragile, silk-like plumage (like Silkies) are visually striking, but they often bring added complexity.

Extreme head feathering can obscure a bird's vision, making them easily startled, difficult to integrate into a flock, and highly vulnerable to predators because they literally cannot see a hawk approaching. Feathered legs collect heavy clumps of mud and waste in wet climates, leading to skin infections or frostbite.

By passing on the rare, visually striking breed in favor of a calm, widely available alternative, you skip the stress of special housing and predator vulnerability. You trade a unique aesthetic for a peaceful, manageable first year.

Keep mixed flocks simple

You might want a mixed flock, getting one buff-colored hen, one black hen, and one white hen, just to tell them apart easily. This is a very practical idea for a beginner.

When mixing breeds, simply try to choose birds with similar size and temperament traits. Mixing a tiny, docile bantam with heavy, dominant standard hens can lead to bullying. Mixing a hyper-active Mediterranean breed with a slow-moving, heavy breed can cause feeding competition. Look for three or four distinct, standard-sized layers that all share a reputation for calm behavior.

Build your short list locally

The goal right now is not to pick the single perfect breed, but to choose from a reasonable pool of locally successful options. To avoid suburban noise and legal complaints, immediately rule out "straight-run" chicks (an unsexed mix of males and females). Always prioritize guaranteed sexed pullets.

From there, talk to two local sources. Whether you are contacting a farm store, a nearby hatchery, or a reputable local breeder, ask them a few specific questions:

- "Which breeds or hybrids do beginners in this area have the easiest time with?"
- "Which birds do you see doing well in our local summer humidity and winter freezes?"
- "Are these chicks guaranteed as sexed pullets?"
- "Are they vaccinated for Marek's disease, and does your facility participate in the National Poultry Improvement Plan (NPIP)?"

Once you know what is locally available, compare those specific options using the traits that matter most to your yard.

For a structured way to evaluate these traits, use Appendix G: Starter flock decision tables. It is designed to walk you through matching your specific climate and space with general categories of sturdy, common hens. By prioritizing local availability and practical traits over rare catalog aesthetics, you stop shopping for a hypothetical, perfect bird. You are ready to choose the real, reliable hens that will actually thrive in your backyard.

Chapter 12: Use starter flock templates for 3, 4, or 6 hens

ALMOST EVERY BEGINNER STARTS with the exact same question: "How many chickens should I get?" It is an understandable question, but it is the wrong one to ask. The better, more practical planning question is: "How many hens can I legally, safely, cleanly, and consistently manage?" If you ask a community of backyard chicken keepers for advice on starting out, you will inevitably hear a joke about "chicken math." It is a popular internet punchline suggesting that no matter how many birds you plan to buy, you will impulsively end up with twice as many.

But when you live in a neighborhood with property lines, barking dogs, and homeowner associations, chicken math is not a joke. It is a cautionary tale that leads directly to suburban nuisance complaints, foul odors, and legal citations. Choosing your flock size must be a concrete planning decision, not a guess at the farm store checkout counter. **Do not max out your legal limit just because you can.** If your city ordinance permits eight hens, that is a legal ceiling, not a starter recommendation. The goal is to start with the smallest flock that fits your family needs, management capacity, and physical space.

Why small is easier for suburban beginners

Think of flock size as the master volume knob for your entire backyard chicken experience. When you increase the number of hens, you turn up the volume on everything. More hens mean hauling heavier bags of feed from your car. They require more pine shavings, produce noticeably more manure, demand a significantly larger physical footprint, and create enough vocal chatter that neighbor relations become critical. Larger flocks also amplify beginner mistakes. If you make a ventilation error or a bedding management mistake with three hens, you have a small, manageable problem to solve. If you make that same mistake with eight or ten hens, you have an odorous, expensive issue that will likely draw flies and neighbor complaints.

You must also understand that rehoming mistakes is rarely an option. Do not buy a large flock assuming you can easily give away the extras if it becomes too much work. Rehoming adult, non-laying, or unwanted hens is surprisingly difficult. Local chicken keepers in your area are often already at their own legal or spatial limits. Furthermore, experienced, responsible flock owners will often refuse to take in adult birds from a stranger's backyard due to the severe biosecurity risk of importing contagious respiratory diseases. Assume you are making a lifelong commitment to the number of birds you bring home.

Verify your space and egg assumptions

Before applying any template, you must ground your expectations in reality. There is no universal "ideal" flock size, but there are physical and biological limitations you cannot cheat.

The space caveat: Do not trust the maximum capacity printed on a prefabricated coop box. A coop marketed "for up to 6 hens" often comfortably holds only three standard-sized birds. A standard baseline assumption is an absolute minimum of 4 square feet of indoor coop floor space and 10 square feet of outdoor run space per standard-sized hen [VERIFY]. However, these are bare minimums, not best practices. The tighter the space, the faster the birds will strip it to bare dirt, the heavier the parasite load becomes, and the more frequently you will have to clean to prevent ammonia buildup.

The egg caveat: When hatcheries advertise that a breed lays "up to 250 eggs a year," that assumes optimal, commercial-style conditions during a hen's youthful peak. In a typical suburban backyard, egg production is rarely linear. During peak season, usually spring and early summer, a healthy young hen might lay five or six eggs a week. But during late summer heatwaves, fall molting, winter darkness, or periods of stress, laying will drastically slow or stop entirely. You will still be paying for feed and managing waste even when the nesting boxes are completely empty.

The 3-hen starter template

Many experienced keepers consider three standard hens to be the most practical starting point for a suburban household. Chickens are social flock animals, and three provides a critical buffer: if you lose one bird to illness or a predator, you are not left with a single, highly stressed solitary chicken.

The household profile: This template works best for cautious beginners, households with smaller yards, or families who travel frequently. It is ideal if your primary goals are low neighbor impact, keeping birds as pets, and occasional fresh eggs.

Space and setup reality: Three hens require a highly manageable footprint. A high-quality 4-by-4-foot coop attached to a secure 4-by-10-foot run easily exceeds minimum space requirements, keeping the birds comfortable and the yard looking tidy.

Feed and cost footprint: A three-hen flock is relatively cheap to maintain. A standard 40-pound bag of layer feed will last them several weeks, making storage simple. Because the daily waste volume is low, a small compost tumbler or dedicated yard-waste bin can easily handle the manure and soiled bedding.

Egg expectations: During peak season in their first two years, expect roughly 12 to 15 eggs per week [VERIFY]. This is plenty for casual weekend breakfasts, but if you want to bake a large cake or host a brunch, you will likely need to supplement with grocery store eggs.

Social dynamics: With only three birds, the pecking order is usually established quickly and quietly. However, it requires you to be highly vigilant about biosecurity; if you ever need to introduce a new single bird later, integrating one newcomer into an established pair can be socially tricky.

The 4-hen starter template

For many suburban households where zoning rules comfortably allow it, four hens represent an excellent middle ground, balancing household utility with manageable chores.

The household profile: This template fits a family that uses eggs regularly, has thoroughly mapped their yard for drainage, and has the budget for a moderately sized, high-quality structure.

Space and setup reality: Stepping up to four hens means paying closer attention to yard capacity. You cross a threshold where run drainage becomes vital. Four hens will quickly turn a small, poorly drained run into an odorous mud pit during a rainy week. You should plan for at least 16 square feet of true indoor coop space and a minimum 40-square-foot outdoor run that is fully roofed to manage the mess.

Feed and cost footprint: Feed consumption is still manageable, but you will notice the increase in bedding costs. Cleaning cannot be delayed; you must stick to a strict weekly routine to prevent flies and odor from reaching the property line.

Egg expectations: Four hens offer a vital egg buffer. If one hen goes broody and aggressively refuses to leave the nest, or if another enters a hard molt, you still have three other birds contributing. During peak season, you can expect roughly 16 to 20 eggs per week [VERIFY], giving a family a much more consistent supply for daily cooking.

Social dynamics: Four is an excellent social number. If the flock splits into pairs while foraging in the run, no bird is left isolated. It also provides slightly more flexibility if you eventually need to integrate new birds, as the flock dynamic is slightly more complex.

The 6-hen starter template

This is a demanding template for a suburban beginner. It requires verified resources, strict legal compliance, and a genuine commitment to daily yard labor.

The household profile: Choose this only if you have a larger lot, no homeowner association, clear municipal rules allowing a flock of this size, and a significant budget. It fits keepers who prioritize consistent egg production for a larger family and are willing to build heavy-duty, permanent infrastructure.

Space and setup reality: Do not assume a standard mail-order coop will work for six hens. Plan on a substantial structure and verify the final dimensions against your local rules, breed size, and the per-bird space standard you are using. Using the minimum baseline discussed above, six standard hens would call for about 24 square feet of indoor space and 60 square feet of run space, with more space preferable where your yard allows. A walk-in run becomes far more practical at this size because you will be spending significant time inside it with a rake and shovel.

Feed and cost footprint: Six standard hens eat rapidly and produce a substantial volume of droppings every single day. You will need bulk, rodent-proof feed storage. You also need a deliberate, high-capacity waste management plan. A small compost tumbler may be overwhelmed; before committing to six hens, verify what your local rules allow and what your household can realistically process. You may need a larger compost setup, such as dedicated bins, or a lawful way to bag and dispose of heavy, soiled bedding on a regular schedule.

Egg expectations: During peak spring laying, six healthy hens will produce 25 to 30 eggs a week [VERIFY]. You will have more eggs than a typical family can eat, meaning you will need a plan for giving them away safely. But remember: you must be prepared to manage the heavy physical footprint of six adult birds in the dead of winter, long after the egg production inevitably drops.

Social dynamics: A six-hen flock is highly active, chatty, and entertaining. The pecking order will be more pronounced, and you will likely see occasional squabbles over the best roosting spots or nesting boxes. They will also absolutely decimate any unprotected landscaping they have access to.

Making your decision

To finalize your flock size, do not rely on a feeling.

Use the table to cross-reference your municipal limit, your usable run space, and your daily time budget. As you evaluate your results, ask yourself two critical questions:

First: If every hen stopped laying from November through February, would you still want to pay to feed and house this specific number of birds? You will be carrying water out to them in the freezing cold long after the novelty has worn off.

Second: Can your setup stay clean with this flock size during a rainy, busy week? It is easy to imagine managing six chickens on a sunny Saturday afternoon when you have free time. It is much harder to clean up after six chickens on a freezing Tuesday evening when you are late getting home from work.

Start small. A successful, clean, stress-free flock of three hens is an incredible joy. An oversized, crowded, odorous flock of six hens is a burden.

Once you have confidently locked in your number of hens, you are almost ready to start looking at breeds. But before you get distracted by beautiful feather patterns and colorful eggshells, there is one specific type of bird you must actively avoid. Choosing the right hens means absolutely nothing if you accidentally bring home a bird that creates guaranteed legal and neighbor problems.

Chapter 13: Skip rooster trouble unless your situation truly fits

If you want fresh eggs, you do not need a rooster. Hens lay eggs entirely on their own as part of their natural ovulation cycle. With proper nutrition, adequate daylight, good health, and a low-stress environment, hens generally produce eggs on a regular schedule. A rooster is not required for egg production, nor will his presence make hens lay more eggs. His only role is fertilization. If your primary goal is a steady supply of breakfast from your backyard, skip the rooster.

If you do end up with an accidental rooster, the eggs your hens lay may be fertile, but that does not mean you will find a developing chick in your breakfast. As long as you collect the eggs promptly each day and store them normally in the refrigerator, they remain exactly the same to eat. Breeding and hatching chicks is a completely different, highly complex project that falls outside the scope of beginner backyard egg production.

For most suburban beginners, adding a rooster changes the entire management burden of a flock. A small group of hens is usually a relatively quiet, low-impact endeavor. Introducing a rooster brings relentless noise, behavioral complications, and legal hurdles that require more experience and physical space to manage.

Before you entertain the idea of keeping a rooster, verify your local laws. Many cities, counties, homeowner's associations, and residential lease agreements explicitly prohibit them, even where backyard hens are warmly permitted. Do not assume you can hide one or bend the rules. Violating ordinances can trigger nuisance complaints that jeopardize your ability to keep chickens at all.

Roosters do not just crow once at a picturesque sunrise. They crow throughout the day. They crow to warn the flock of perceived danger, assert dominance, respond to a passing truck, or simply communicate with the hens. Assuming neighbors will eventually get used to the sound is a risky bet that usually ends in conflict.

Because their primary instinct is to protect the flock, some roosters become highly territorial. In a tight suburban yard, that protective instinct might be aggressively directed at a visiting child, your family dog, or even you when you walk out to fill the waterer.

Some keepers mistakenly believe a rooster will solve their predator problems. While a rooster will certainly sound the alarm if he spots danger and might try to defend the hens, he is not a substitute for secure housing. He is no match for a determined raccoon, a hungry fox, or a stray dog. Your real predator protection comes from secure coop design, heavy-gauge hardware cloth, and locking the flock up every night, not from a rooster.

Even if you are committed to a hen-only flock, you must prepare for the possibility of an accidental rooster. This happens most often when beginners buy young chicks. Sorting newly hatched chicks by male and female is called "sexing." It is a highly skilled visual assessment, but it is not a perfect science. Many reputable hatcheries and farm stores guarantee around 90 percent accuracy when selling pullets, the industry term for young female chickens. That means one out of every ten chicks sold as a female might turn out to be a male.

You can drastically lower this risk by starting with older pullets or adult hens. By the time a chicken is fully feathered and approaching laying age, its sex is visually obvious. However, depending on the breed and the reliability of your source, even buying older birds carries a slight risk. Late-blooming roosters or misidentified breeds from unverified backyard sellers can occasionally surprise you.

Consider a common suburban scenario: a family orders a small batch of chicks. Around fourteen to sixteen weeks of age, one of the birds stretches its neck and lets out a raspy crow. Because the family assumed they only had hens, they have no idea what to do next.

If a chick turns out to be male, pause and assess the situation. First, verify your rules. If roosters are illegal, or if your yard cannot safely accommodate a territorial bird, he cannot stay.

If you must remove the bird, you will quickly discover that ethical rehoming is incredibly difficult. Because roughly half of all chickens hatched are male, and very few residential areas allow them, the market is flooded with unwanted roosters. Local farmers usually already have breeding roosters and do not want to introduce outside disease or competition to their flocks. Many animal shelters and farm rescues operate at maximum capacity and turn poultry away.

Never dump or release an unwanted rooster in a park, on a roadside, or into the wild. Abandoning a domestic animal may be illegal and may be treated as animal cruelty or neglect depending on state and local law, and it exposes the bird to predators, weather, hunger, traffic, and disease. Use lawful rehoming options such as animal control, a poultry rescue, a sanctuary, a veterinarian, or your Extension office.

If you buy young chicks, ask your source one vital question before finalizing your order: "What happens if a chick turns out to be male?" Some local breeders and farm stores have a take-back policy for accidental roosters. Others will explicitly tell you that once the bird leaves the store, you are entirely responsible for it.

Before you bring a single bird home, formalize your safety net using the [ASSET REQUIRED: Rooster contingency checklist]. Determine your local rescue contacts, your seller's return policy, and your exact steps if a crow echoes from the coop, utilizing the [ASSET REQUIRED: Surprise-rooster response plan].

A hen-only flock is not a compromise. It is the simplest, quietest, and most neighbor-friendly way to enjoy backyard chickens. By planning for hens and preparing for the unexpected, you set your suburban flock up for success.

Part 4: Bring birds home and manage the first 90 days

You have navigated property lines, verified local zoning ordinances, deciphered homeowner association rules, and chosen the quietest, most secure corner of your yard for the coop. You have mapped out your property drainage, confirmed your legal setbacks, and had honest conversations with your neighbors. The theoretical work is finished. Now comes the pivot point where your project becomes physical.

Bringing backyard chickens home marks a major shift from project planning to daily livestock responsibility. Whether you are driving to the post office to pick up a peeping cardboard box of day-old chicks or returning from a local farm with a ventilated dog crate full of nervous pullets, the drive home changes everything. The chapters in this section are designed to guide you through that exact transition.

The overriding theme of Part 4 is a non-negotiable rule of livestock management: finish the system before the birds are in the car. This part opener serves as your roadmap for the next three months. It frames the decisions, habits, and household rules you must establish to ensure your first flock thrives.

The "finish the system" rule

Beginners often feel a sudden wave of anxiety as they carry their new birds across the lawn. You look at these small, fragile animals, and a dozen questions rush into your mind. Will they be too cold tonight? Is the feeder in the right spot? What if a raccoon tests the latches? This nervous energy is completely normal. Because you care enough to worry, you are already building an observant mindset.

To safely ease that anxiety, you must have every physical element of your setup tested and running before pickup day. You cannot leave shivering chicks sitting on your kitchen counter while you search the garage for a heat source. You cannot leave stressed pullets waiting in a hot car while you try to figure out how to assemble a galvanized waterer or realize you forgot to buy the right starter feed.

"Finishing the system" means the indoor brooder or outdoor coop is fully assembled. If you are starting with chicks, the heat source is plugged in, properly secured against fire risks, and holding the correct temperature. The bedding is laid down. The waterers are full and waiting. The bulk feed is secured in heavy-duty, rodent-proof bins. If the birds are going outside, the predator-resistant hardware cloth is fully attached, the digging aprons are trenched, and every single latch has been tested. The habitat must be waiting for the birds, not being built around them.

Choose your ninety-day path

The first ninety days of chicken keeping will look drastically different depending on the age of the birds you decided to bring home. As you move through the upcoming chapters, you will navigate the specific path that aligns with your acquisition choice.

Path 1: Day-old chicks If you chose to start with day-old chicks, your first ninety days are heavily front-loaded with indoor management. You are essentially acting as their mother hen. The first thirty days will revolve around brooder safety, managing a heat source, monitoring for rapid growth, and dealing with an escalating amount of fine bedding dust in your house or garage. Around week six to eight, depending on outdoor weather and their feathering progress, you will execute a carefully planned transition to the outdoor coop. Your ninety-day milestone will likely coincide with them fully integrating into outdoor life, though they will still be weeks or months away from laying their first eggs.

Path 2: Started pullets If you chose started pullets, teenage birds that are fully feathered but not yet laying, your ninety days bypass the indoor brooder phase entirely. These birds go directly into your secure outdoor coop. Your initial focus will be on teaching them where home is, getting them accustomed to their enclosed run, and establishing your daily outdoor maintenance routines. Sometime between day thirty and day ninety, depending on their exact age upon arrival and the season, you will likely transition into daily egg collection.

Path 3: Adult hens If you are bringing home a flock of mature, currently laying hens, your daily chore routine starts immediately at full capacity. You bypass both the brooder setup and the waiting period for eggs. Your ninety-day window focuses heavily on helping older birds overcome the stress of relocation, adjusting to their new suburban environment, and carefully monitoring them for any underlying health issues that might emerge after the stress of moving.

The three phases of your new routine

Regardless of which path you choose, the first ninety days represent a distinct learning curve. We will break this curve down into three manageable phases to help you build competence without feeling overwhelmed. Rather than listing every single daily task here, you will find detailed, printable daily and weekly tracking calendars in the companion PDF tool system.

Phase 1: The first week of arrival and observation Your first week is about calm observation. When birds arrive, they are experiencing significant stress. Your job is to provide water, warmth, and security, and then step back. Limit handling, keep eager children sitting quietly, and maintain strict physical separation from curious family dogs. Sitting quietly teaches you their baseline behavior. By learning what normal looks like, you train your eyes to instantly notice when a bird is acting sick or injured later.

Phase 2: Days 8 to 30 for building the habit Once the arrival excitement settles, you will focus on building the repeatable routines that keep your flock legal, safe, clean, contained, and neighbor-friendly. You will build a five-minute morning checklist: unlocking the coop, checking the water level, and inspecting the perimeter for predator digging. You will also establish your weekly cleaning rhythm to prevent ammonia odors from drifting across the fence line.

Phase 3: Days 31 to 90 for transition and maturity The final stretch is marked by transition. Chicks move outside, pullets mature into laying hens, and your skills solidify. By day ninety, your daily checks will feel like second nature, and you will conduct your first major system review to ensure your flock remains a sustainable part of your household.

Moving forward

The transition from an empty yard to an active backyard flock is thrilling, noisy, and occasionally messy. It requires patience, a willingness to get your hands dirty, and a commitment to showing up every single day, regardless of the weather. The groundwork you laid in the previous sections secured your legal standing and property layout. Now, you must finalize your household habits. Your next step is to lock down your supplies and set up a safe, fully tested environment. Let the systems you have built do the heavy lifting, so when the birds finally arrive, you can focus entirely on welcoming them home.

Chapter 14: Get supplies, records, and routines ready before arrival

PICTURE BRINGING BACKYARD CHICKENS home in two different ways.

In the first scenario, you fall in love with fluffy chicks at a farm store and buy on impulse. On the drive home, panic sets in. You have no heat source, no pine shavings, and nowhere to put them. Your dogs wait at the door. You spend the evening running back to the store, scrambling to build a makeshift garage brooder, and fighting off pets.

In the second scenario, you drive home and walk into a quiet house. You carry the transport box to a secure, assembled habitat. The heat source is tested. The water is filled. Feed waits in a secure metal bin. Dogs are inside, and your family knows the arrival rules. You simply open the box and watch the birds settle in.

Bringing hens home is like bringing home any dependent animal. Physical supplies and daily routines must precede the living creature. Arrival day should be uneventful because the system is already built.

The before-arrival rule and biosecurity prep

Do not bring birds home until you can pass a basic pre-flight inspection. An empty wooden box sitting in the grass is not a functioning habitat.

Before purchasing supplies, establish household biosecurity basics to prevent spreading infectious disease to your birds, and from your birds to your family. Biosecurity is the practice of limiting disease transmission through simple, consistent physical barriers. Purchase a dedicated pair of "coop boots" worn only inside the run, never in the house. Set up a handwashing station or designate an outdoor sink for use after handling chickens. If picking up adult birds, wash and sanitize the transport crate beforehand. Do not track farm-store or breeder-yard germs into your vehicle or home.

Checklist-driven supplies by pathway

Do not assume a farm store clerk will tell you everything you need. Retail workers are exceptionally busy during chick season. It is your job to have the habitat ready. Use the following checklists to verify supplies based on your acquisition pathway.

Core supplies for all setups

- **Secure feed storage:** A galvanized steel trash can with a tight-fitting lid to prevent rodent infestations.
- **Cleaning tools:** A stiff-bristled brush, a scraper, and a designated coop bucket.
- **Dedicated footwear:** Rubber boots for coop use only.
- **Quarantine bin:** A clean wire dog crate ready in case you need to isolate an injured bird.

Supplies for day-old chicks

- **Brooder enclosure:** A high-sided, draft-free container (such as a plastic stock tank).
- **Heat source:** A commercial brooder heat plate or a securely mounted, wire-guarded heat lamp.
- **Thermometer:** To monitor temperature at chick level.
- **Feed:** Commercial chick starter feed appropriate for their age; ask the hatchery, seller, veterinarian, or feed label whether medicated or unmedicated starter is appropriate for your situation.
- **Waterer:** A shallow, chick-specific waterer to prevent drowning.
- **Bedding:** Bulk large-flake pine shavings. Avoid cedar shavings or slippery newspaper.

Supplies for juvenile pullets

- **Housing:** A secure, predator-resistant coop and run.
- **Feed:** Grower or developer feed formulated for their specific age.
- **Feeders and waterers:** Adult-sized equipment, elevated to back height to keep bedding out.
- **Bedding:** Pine shavings, hemp, or coarse sand, depending on your setup.

Supplies for adult laying hens

- **Housing:** A secure coop and run with appropriate roosts.
- **Feed:** Commercial layer feed appropriate for actively laying hens.
- **Calcium supplement:** A separate dish of crushed oyster shell provided free-choice.
- **Nesting box supplies:** Clean, soft nesting material like excelsior pads or fresh pine shavings.
- **Transport carrier:** A sturdy, well-ventilated plastic dog crate.

Records to create

Recordkeeping is often an afterthought, but details fade quickly. Set up a basic notebook or digital folder with these exact fields before the birds arrive, and keep it easily accessible:

- **Source information:** Name, phone number, and address of the breeder or store.
- **Bird demographics:** Hatch date or exact age, breed, and headcount.
- **Vaccination records:** Verify and log whether they received hatchery vaccinations for Marek's disease or Coccidiosis.
- **Feed history:** The exact brand and type of feed the seller used, ensuring a smooth transition.
- **Health notes:** A dedicated space for first-day observations.

Emergency contacts

A dangerous myth is that you only need emergency contacts after something goes wrong. Waiting until a hen is sick causes unnecessary stress. Log these resources in your binder now:

- **Veterinarian:** Locate a poultry-savvy avian veterinarian in your area, if available.
- **Agricultural Extension:** Find the number for your county extension office for localized, science-based advice.
- **State authorities:** Identify your state's diagnostic lab or animal-health contact for suspected contagious diseases or sudden flock death.
- **Local compliance:** Have the direct number for your local animal control or zoning office.

Use Appendix O: Sick-bird help packet to log these numbers immediately. It contains a help-call worksheet and a blank emergency contacts list designed to be filled out before arrival.

Household routines and boundaries

A major point of failure occurs when a family assumes coop chores will simply get done, or expects children to handle them entirely.

There must be an assigned adult owner of responsibility. This adult ensures the coop is locked at night, water is fresh, and birds are healthy, even when a child has evening soccer practice. Decide who opens the coop in the morning and who checks the water before work. Establish a firm weekly cleaning schedule and a vacation backup plan with simple written instructions for your sitter, rather than relying on verbal promises.

Equally important are physical boundaries. Decide exactly where children are allowed to interact with the birds, and enforce strict handwashing rules. Establish a clear boundary for pets. Decide in advance where dogs will be during arrival and daily chicken care so they are not staring aggressively through the run fence. Review *Appendix I: Egg hygiene and household boundaries packet* with your family, ensuring everyone understands the child, dog, and guest plans before the first bird sets foot on your property.

Pickup and transport basics

The drive home is a vulnerable time. Transport is inherently stressful, and poor conditions can cause overheating, chilling, or injury.

If picking up adult hens or pullets, use a rigid, well-ventilated pet carrier lined with pine shavings or a puppy pad for grip. Avoid transporting birds loose in a vehicle; a secure carrier reduces escape, injury, and sanitation risks. Day-old chicks usually arrive in a small, ventilated cardboard box. Leave them in this box until you are safely home.

Follow these transport rules:

- **Direct trips only:** Go straight home. Do not run errands or leave birds in a parked car.
- **Climate control:** Adult birds are highly susceptible to heat stress; keep your vehicle cool and well-ventilated without pointing air conditioning vents directly at them. Chicks chill easily; keep the car comfortably warm and block cold drafts.
- **Smooth driving:** Avoid sudden braking or sharp turns that can throw birds against the side of the carrier.

Arrival-day schedule

When supplies are stocked, records are prepped, contacts are logged, and the family knows their jobs, you are ready for arrival day. Focus on a quiet, low-stress transition.

- **Step 1: Secure the area.** Lock dogs inside the house. Ask children to observe quietly from a distance. Limit outside visitors.
- **Step 2: Check the environment.** Ensure the brooder heat source is stable or the adult coop is completely secure.
- **Step 3: Prepare food and water.** Have waterers filled with room-temperature water. For day-old chicks, gently dip the tip of each chick's beak into the waterer as you place them in the brooder so they know exactly where it is.
- **Step 4: Release and step back.** Gently place the birds into their habitat. Do not crowd them.

First-week observation plan

Once the birds are in their new habitat, your primary job shifts to observation. Do not handle them excessively during the first week. Spend time watching them to establish a baseline for normal behavior.

Pull out your record book and jot down your first observations. A good first-day observation line for chicks should look like this:

"Date: May 12. Time: 4:00 PM. All six chicks drinking independently. Active and exploring. Sleeping scattered around the edge of the heat plate. No huddling or loud distress peeping. Droppings look normal."

For adult hens, a normal observation might read:

"Date: September 4. Time: 2:00 PM. Four pullets placed in run. Nervous but walking normally. Two have found the waterer. No signs of wheezing, lethargy, or injury."

Taking quiet notes centers you in your role as an observant flock manager. With your supplies stocked, systems organized, and routines assigned, you have built the infrastructure of a successful backyard flock. If you are starting with chicks, the next step is making sure the brooder itself is safe and ready.

Chapter 15: If you start with chicks, set up the brooder with care

PICTURE BRINGING BACKYARD CHICKENS home in two different ways. In the first scenario, you fall in love with fluffy chicks at a farm store and buy on impulse. On the drive home, panic sets in. You have no heat source, no pine shavings, and nowhere to put them. Your dogs wait at the door. You spend the evening running back to the store, scrambling to build a makeshift garage brooder, and fighting off pets. In the second scenario, you drive home and walk into a quiet house. You carry the transport box to a secure, assembled habitat. The heat source is tested. The water is filled. Feed waits in a secure metal bin. Dogs are inside, and your family knows the arrival rules. You simply open the box and watch the birds settle in.

Bringing hens home is like bringing home any dependent animal. Physical supplies and daily routines must precede the living creature. Arrival day should be uneventful because the system is already built. Do not bring birds home until you can pass a basic pre-flight inspection. An empty wooden box sitting in the grass is not a functioning habitat. This chapter is designed to help you prepare the space safely before you ever put a carrier of peeping birds in your car. Reference Appendix N: Boundary Plan

Who should choose chicks

Before building a brooder, confirm your household is ready for baby birds. Raising chicks requires a specific commitment that differs wildly from buying adult birds. First, it requires daily, sometimes hourly, monitoring. You cannot leave day-old chicks alone for a weekend. They need someone present to check their water, monitor their behavior, and ensure their heat source is functioning properly. Second, you must be able to manage safety boundaries. If you live in a home with highly active hunting dogs, curious toddlers, and no spare utility room, securing a warm, dusty brooder will be challenging.

Many aspiring keepers review the requirements for heat safety and dog barriers and realize a brooder does not fit their home. They buy started pullets instead, older, fully feathered birds ready to move directly into the outdoor coop. Started pullets are a highly practical choice, not a failure. Finally, have a rooster contingency plan. Even professionally sexed pullets are not guaranteed; ask the hatchery about its accuracy rate and policy. If your local rules restrict or ban roosters, confirm the current ordinance with the appropriate local office and make a legal, humane rehoming plan for a crowing bird before you buy chicks.

Where to place the brooder

Location is your first major decision. Because chicks are small, the instinct is to keep them close. However, the Centers for Disease Control and Prevention (CDC) advises that backyard poultry should not be allowed inside the home for any reason, especially in areas where food or drinks are prepared, served, or stored, such as kitchens, dining areas, pantries, and other food-preparation or food-storage areas. Healthy-looking chicks can carry bacteria, including Salmonella and Campylobacter, in their droppings and on their feathers.

Chicks also generate "chick dust," a thick, fine particulate made of dried droppings, bedding, dander, and feed. You need a location protected from children and pets, free from drafts, but properly ventilated. A dedicated utility outbuilding, detached workspace, or heated garage can work well, provided the space is entirely free from vehicle exhaust, chemical fumes, and unsafe temperature swings.

Use this quick location checklist when choosing a spot:

- Absolutely no kitchens, bathrooms, or food-prep areas.
- Washable, easily cleaned surroundings.
- Protected by a closed door from pets and young children.
- Easy to check frequently.
- Draft-free, but ventilated.
- Safe clearance for a heat source and electrical cords.

Brooder basics and feed stages

A brooder acts as an artificial mother and a nursery. It does not need to be expensive, but it must be functional. A minimum setup includes:

A safe enclosure: Many keepers use deep galvanized metal tubs, heavy-duty plastic storage totes, or commercially manufactured brooders. Walls must be tall enough to contain jumping chicks and sturdy enough that a bumping dog cannot knock the enclosure over. It must have a secure, breathable top, such as heavy hardware cloth fastened over the opening, to keep jumping chicks in and household hazards out.

Appropriate bedding: Agricultural Extension services generally recommend large-flake pine shavings. Avoid cedar shavings, which are widely reported to irritate poultry respiratory systems. Avoid slippery surfaces like plain newspaper, which can cause chicks to develop permanent leg deformities.

Specialized waterer: Farm stores sell shallow, chick-specific waterers designed to prevent the birds from stepping in. Standard agricultural safety guidance warns that chicks can easily drown in deep dishes or household bowls.

Chick starter feed: Offer constant access to a fresh, age-appropriate chick ration to support their rapid growth. Do not feed adult layer pellets to baby chicks. Layer feed contains calcium levels meant for hens actively producing eggshells; it is not formulated for chicks and can harm them. Keep chicks on a starter or grower formulation until they reach the age recommended by the feed manufacturer.

Heat, electrical, and fire safety

Chicks hatch unable to regulate their own body temperature. For their first several weeks, they rely entirely on an external heat source. Heat lamps featuring an infrared bulb are the traditional method, but they carry extreme fire risk. They get dangerously hot and are a documented source of devastating barn and garage fires. If you choose to use a heat lamp, you must consult qualified fire-safety and Extension guidelines.

Keep lamps clear of bedding, cardboard, plastics, and other combustibles. Inspect all cords for damage before use. Most importantly, do not rely on a heat lamp's built-in spring clamp as the only support. Suspend the lamp from a secure chain, cable, or manufacturer-approved hanger, and use the clamp only as a secondary positioning aid if the manufacturer allows it. Many keepers now choose radiant brooder plates to avoid suspending a hot bulb. These devices sit on adjustable legs, radiating warmth downward and allowing chicks to duck underneath, simulating a mother hen. While they reduce certain risks, they are still electrical appliances and must be treated with respect.

No matter which heat source you choose, you must:

- Plug equipment only into properly installed outlets that are appropriate for the location and the equipment; for garages, utility rooms, damp areas, or outdoor/coop spaces, have a qualified electrician confirm GFCI and local-code requirements.
- Use only properly rated, listed brooder-heating equipment according to the manufacturer's instructions.
- Keep electrical cords secured and out of the brooder enclosure.
- Secure heat lamps with chain, wire, or a fixed mount so they cannot fall, and never hang a heat lamp by its cord.
- Keep heat sources away from bedding and other combustibles.
- Do not use extension cords in the coop or as permanent wiring.
- Have any new outlets, outdoor or coop circuits, or permanent wiring installed or inspected by a qualified electrician and follow local code.
- Follow all manufacturer instructions to the letter.

Set up and test your heat equipment for a full 24 hours before the chicks arrive. Consult your hatchery or local Extension office for a precise weekly temperature schedule, which typically starts high and steps down gradually as feathers develop.

Daily observation and health limits

Once the chicks are in the brooder, your primary job is observation. Keeping a daily log helps you notice subtle changes before they become emergencies. Watch their posture to gauge the temperature. Extension guidance notes that chicks will tell you if they are comfortable. If they are huddled in a tight, noisy pile directly under the heat source, they are too cold. If they are pressed against the far edges of the brooder, panting with their beaks open, they are dangerously hot. If they are scattered loosely, eating, sleeping, and peeping softly, the temperature is generally right.

Beginners must also monitor droppings. A common, potentially fatal condition in young chicks is "pasty butt," where droppings stick to and block the vent area. If you notice a blocked vent, do not attempt aggressive home remedies and do not apply random household oils. You may hold the chick securely and gently soften the dried droppings with a warm, damp cloth until they wipe away. Never pull on dried feces, as this can tear the chick's fragile skin. If pasty butt persists, if breathing appears labored, or if multiple chicks act lethargic, do not turn to internet forums for medical treatments. Contact a poultry veterinarian or your local Extension office for expert guidance.

Household boundaries and CDC hygiene

Chicks are undeniably appealing, but they require strict household boundaries. Every interaction between a child and a chick must have direct adult supervision. You must establish clear hygiene rules based on CDC guidelines for backyard poultry. According to the CDC, children under five years old, adults over 65, pregnant people, and anyone with a weakened immune system should not handle or touch chicks, as they are at a higher risk for severe illness from Salmonella.

For the rest of the household, the rules are simple and non-negotiable:

- No kissing, snuggling, or holding chicks near your face.
- No eating or drinking near the brooder.
- After handling chicks, touching the brooder or enclosure, changing water, or handling any feed or water containers or other brooder equipment, wash hands right away with soap and clean, running water, warm or cold, for at least 20 seconds. If soap and water are not readily available, use a hand sanitizer with at least 60% alcohol until hands can be washed.

Pets and brooder safety

Do not assume your household pets will understand that the chicks belong to you. To a dog or cat, a fluttering, peeping chick sounds and moves exactly like prey. Dogs and cats should not have unsupervised access to the room where the brooder is kept. "The dog just wants to look" is a dangerous mindset that routinely ends in tragedy. You cannot rely on verbal commands or training guarantees when dealing with high-value prey drive. You must use physical barriers. The brooder room requires a secure, latched door, and the enclosure itself needs a heavy, fastened hardware-cloth lid that allows ventilation while keeping paws out. Treat the brooder as a restricted zone.

Planning the transition outside

You might look at the tiny birds in your brooder and think you have plenty of time to build their permanent coop. You do not. Chicks grow at an astonishing rate. Within weeks, they will be testing their wings, kicking pine shavings over the walls, and producing significantly more waste. Your outdoor coop and run must be fully built, legally placed, predator-resistant, and ready before you bring chicks home. Moving them outside depends on their feather development, their age, their health, and your local weather. In many climates, chicks are not ready to live outside without supplemental heat until they are well feathered and nighttime temperatures are appropriate for their age and condition.

Avoid relying on universal timelines, as moving day varies widely between a warm southern climate and an environment with late-spring snowstorms. Starting with chicks is highly rewarding when you prepare properly. By securing the heat source, managing the dust, enforcing household hygiene, and building the outdoor coop in advance, you remove panic from the process. With the brooder safely established, you can focus on bringing your birds home and settling them into a calm, prepared environment.

Chapter 16: If you start with pullets or hens, settle them in calmly

Some beginners assume older pullets and adult hens are the "plug-and-play" option for backyard chicken keeping. Bypassing the fragile chick phase means skipping heat lamps, indoor brooders, and the daily anxiety of temperature fluctuations. You might assume you can simply bring a box of fully feathered birds home, open it in the backyard, and watch them march toward the nesting boxes to lay your morning breakfast.

But older birds are not plug-and-play. Moving disrupts their environment, flock dynamics, and baseline sense of safety. Furthermore, adult birds carry a hidden variable that day-old chicks usually do not: a life history. Unlike chicks hatched in a controlled environment, adult hens and started pullets have lived outdoors, interacted with wild birds, and been exposed to local pathogens. If handled casually, adult-bird acquisition can become an easy way to import disease directly into your backyard.

Bringing hens home safely requires managed logistics. The physical supplies, disease-prevention strategies, and daily routines must precede the living creature.

Seller screening and biosecurity

Careful observation assumes you started with healthy birds, which means vetting your source before you ever leave your driveway. Not all rehomed hens are healthy, and adult birds from uncertain sources require careful questioning.

Document the flock's history. Ask the seller for the birds' known breeds, exact ages, current daily feed formulation, and any available vaccination records. Most importantly, ask if the seller's flock has experienced any recent respiratory illnesses, sudden deaths, or treatments. Keep this information in your permanent flock records; if your birds get sick later, a veterinarian will need this history.

If a seller advertises their flock as NPIP-certified (National Poultry Improvement Plan), ask for current documentation and confirm what that status covers at the time of sale. NPIP participation is valuable, but it should not be treated as a blanket health guarantee; it does not prove the birds are free from every common poultry parasite, respiratory virus, or bacterial infection.

The mixed-source disease risk

Because this is your first flock, you should acquire all of your birds at the exact same time from a single, reputable source.

One of the biggest biosecurity risks a beginner can take is building a "mixed-source" flock on day one. It is tempting to buy two hens from a local farm swap, three from a neighbor, and one from an online classified ad, combining them all in your new coop. Do not do this.

Different farms have different baseline pathogens and different flock immunities. The birds from Farm A might be asymptomatic carriers of a respiratory virus they fought off months ago. The birds from Farm B might have never encountered that virus. When you mix them together under the high stress of a move, the Farm A birds shed the virus, and the Farm B birds can become seriously ill. Starting with a single-source flock minimizes this collision of pathogens and makes tracking health history much easier.

Transporting older birds

Transporting fully grown birds requires specific gear and strict attention to temperature. A flimsy cardboard box with three pencil holes punched in the top is a suffocation risk.

Secure heavy-duty, well-ventilated transport boxes explicitly designed for poultry, or use hard-sided plastic dog and cat crates. Line the bottom of every carrier with a thick towel or a layer of pine shavings so the birds have traction. Sliding around on a smooth plastic floor causes panic and potential leg injuries.

Manage heat and cold stress aggressively. Poultry run hot, and an enclosed box inside a warm car can become fatal in minutes. If you are picking up birds in the summer, travel in the early morning or late evening to avoid the heat of the day. Keep the car cabin cool and well-ventilated, but do not point air conditioning vents directly into the carriers, which can cause severe chilling.

Keep the drive direct. Go straight from the seller to your home. Do not run errands, and do not open the carrier in the driveway to peek at the birds or show them to your children.

Arrival day logistics

Do not bring birds home until you can pass a pre-flight inspection. An empty wooden box sitting in the grass is not a functioning habitat. Your coop and run must be fully assembled, secured against predators, lined with bedding, and stocked with feed and water.

When you arrive home, follow this sequence to get the flock settled:

- Bring the carriers directly to the prepared, secured run.
- Place the carriers on the ground inside the run and open the doors or flaps.
- Step back. Let the birds step out on their own time. Do not reach in and drag them out by their wings or legs.
- Confirm they can easily find and access the feed and water.
- Walk away. Keep family, dogs, and neighbors out of the yard.
- Perform a quiet check at dusk to ensure the birds have moved inside the coop to roost and that all doors and latches are securely locked for the night.

Do not let your new flock free-range immediately. They do not yet know the coop is their home base. If they are startled by a dog barking next door, their instinct will be to scatter. Frightened pullets can fly over fences, squeeze through surprisingly small gaps, or bolt farther than you expect. Keep them confined to the locked coop and run for at least the first week so they map the exact location of their food, water, and shelter.

Set strict household boundaries. It is natural for children and neighbors to want to meet the flock immediately, but the birds need quiet to decompress. Make sure everyone understands the first week is for observation, not handling or socializing.

Observation and quarantine realities

If you are starting your very first flock in an empty yard, the birds can go straight into their new, permanent coop. If you already have chickens, do not place new birds directly into the established flock. Quarantine new or returning birds in a separate area for at least 30 days, watch them daily for illness or parasites, and only then begin a gradual, supervised introduction.

Use that separation period to observe the new arrivals before they share waterers, equipment, or close contact with your original birds.

However, avoid overpromising quarantine effectiveness. While a 30-day isolation period is a critical best practice to catch active illnesses, it is not a magical shield. Some diseases can involve asymptomatic carriers. A bird may appear perfectly healthy during quarantine, only to shed a pathogen later when stressed by extreme weather or molting. This is why biosecurity, seller screening, and single-source purchasing remain important defenses.

First-week observation

During the first seven days, your primary job is distant observation. Moving triggers a reshuffling of the pecking order and suppresses the immune system. Use the [ASSET REQUIRED: First-week settling-in tracker] to establish a baseline of normal behavior. Check on the flock daily and observe:

- **Eating and drinking:** They may eat lightly on the first day, but you should soon see them actively scratching at the feeder and drinking regularly.
- **Posture:** A healthy bird explores the run and interacts with others. A bird standing hunched in a corner with its feathers puffed out for hours is a serious concern.
- **Breathing:** Breathing should be silent and relaxed, without wheezing, clicking, or open-mouthed gasping (unless they are panting lightly to cool down on a hot day).
- **Droppings:** The stress of moving can cause temporary loose stools. If watery or discolored droppings persist beyond a day or two, or are paired with lethargy, take note.
- **Laying changes:** Stress halts egg production. A pullet about to lay her first egg, or an adult hen in prime production, will likely stop laying after a move. Give them time to feel safe; do not confuse moving stress with illness, but do not ignore physical symptoms.

When to seek help

Even with careful sourcing, the stress of a move can occasionally cause underlying health issues to flare up. You need an escalation plan before it happens. If one bird appears sick, keep the bird calm and contained. Minimize handling, record the specific symptoms, note when they started, and contact an appropriate poultry professional.

Watch for immediate red flags. If you observe severe lethargy, a bird refusing to drink, or unusual respiratory signs like bubbling eyes, swollen sinuses, or heavy coughing, do not wait to see if it resolves. If multiple birds act abnormal simultaneously, or if you experience a sudden, unexplained death in the flock, seek professional guidance immediately. Contact a local poultry veterinarian, your university Extension office, or your state diagnostic lab. Do not rely on internet forums or social media groups to diagnose serious respiratory symptoms or sudden death.

Crucially, do not transport a sick or dead bird to a veterinary clinic, Extension office, or farm store without explicit instructions to do so. If your flock may be dealing with a suspected or confirmed highly contagious disease such as Highly Pathogenic Avian Influenza (HPAI) or virulent Newcastle disease, loading a sick or dead bird into your car and driving it across town can risk spreading the disease to other flocks and agricultural centers. Call your diagnostic lab or state veterinarian first, describe the symptoms, and follow their current protocol.

By preparing your habitat entirely in advance, insisting on strict seller biosecurity, and giving your new flock a quiet, strictly managed arrival, you replace chaos with routine. Older birds require respect, distance, and observation to settle in safely. Ensure the system is ready before the birds ever arrive.

Chapter 17: Build the daily and weekly routine that prevents most problems

Many beginners imagine that keeping chickens requires a heroic level of effort. They picture themselves waking before dawn, hauling heavy bags of feed in the freezing rain, scrubbing coops with a wire brush, and constantly managing farm-level crises. But successful suburban chicken keeping is not about heroic effort. It is about ordinary, predictable consistency. You might have heard the phrase that chickens are "low maintenance." This is an incomplete truth. A backyard flock is predictable maintenance, but it still requires daily attention.

If you wait for problems to announce themselves, you often end up dealing with a sick bird, a ruined run, or a neighbor complaining about an unmanageable odor. Routine is your flock's safety net. When you follow a predictable pattern, the baseline of what looks, sounds, and smells normal becomes deeply ingrained in your mind. This makes the abnormal instantly obvious. To keep your flock clean, contained, and neighbor-friendly, you need an inspectable routine.

Time-cost realism

Before breaking down the tasks, set a realistic baseline for the time involved. A well-designed, securely built setup requires about ten minutes of daily active work, twenty to thirty minutes of weekly maintenance, and one to two hours of deeper monthly inspection. If you find yourself spending hours every weekend managing manure, fighting mud, or hauling water, either your setup or your daily routine needs adjusting. Skipping a ten-second daily spot-cleaning often creates a thirty-minute chore by the weekend. Appendix L includes daily, weekly, and monthly routine schedules.

The daily scan

Daily care should not feel like an overwhelming chore. In fact, your morning check can usually be accomplished through a quick, repeatable "two-minute daily scan." Use this exact sequence when you visit the coop each morning:

- **Feed:** Is there enough feed in the dispenser to last the day? Is it dry, unsoiled, and flowing freely?
- **Water:** Is it full, clean, and unfrozen? Chickens require reliable daily access to clean water during all waking hours.
- **Visual check:** Are the birds moving normally, scratching, and eating? Is anyone hiding in a corner, holding their tail unusually low, or refusing to leave the coop?
- **Egg collection:** If your hens are laying, collect the eggs daily. Prompt collection prevents eggs from freezing in winter, baking in summer, or being stepped on and eaten by the flock.
- **Bedding spot-check:** Look for a tipped waterer, a feeder scratched empty, or a sudden accumulation of damp bedding under the roost. A quick scoop with a hand trowel right now takes ten seconds and prevents ammonia buildup.
- **Gate and latch check:** Ensure nothing was damaged overnight. As you leave, physically verify that the run gates latch securely behind you.

- **Weather check:** Anticipate the day's conditions. Will you need to add a shade block for an afternoon heat wave, or safely plug in a heated waterer (following all manufacturer instructions) ahead of a freeze?
- **Predator check:** Scan the perimeter for new digging attempts, disturbed soil around the run apron, or claw marks on the wood.

As you do this scan, observe the birds themselves. Most healthy hens usually stretch, hop out of the coop, and immediately begin scratching and pecking. If a hen stays huddled on the roost or refuses to come out for a morning treat, treat it as a reason to pause and observe her more closely.

Minimum viable care

Some mornings turn chaotic. You oversleep, a child is sick, or a severe thunderstorm hits exactly when you need to leave for work. On these days, rely on a minimum viable routine. If you can do nothing else, ensure these non-negotiables happen: verify the water is full and unfrozen, check the feed levels, count the birds to make sure no one is visibly injured or distressed, and secure the doors.

The evening lockup

The evening routine is your primary defense against predators. Suburban nights are busy, and the temptation to assume the chickens put themselves to bed is strong. While chickens do naturally return to the roost at dusk, your job is to secure the perimeter. Even if you use an automatic coop door, you still need to perform periodic human checks to ensure it closed properly on its track and no bird was locked outside. Count heads to ensure all your hens are accounted for. Check the latches on the coop door and run. Ensure they are locked tight against raccoons, opossums, or neighborhood dogs. Take a quick look around the base of the run with a flashlight to see if anything has tried to dig under the wire during the day. This routine takes only a minute or two, but it is one of the most vital habits you will build to keep your birds alive.

Weekly tasks

Once a week, step slightly beyond the daily scan to reset the setup for the days ahead. Do not fall for the misconception that cleaning can wait until the coop smells. A noticeable ammonia or manure odor means the setup is already failing and irritating your birds' sensitive respiratory systems.

- **Cleaning:** Depending on your management system, remove soiled bedding and refresh the coop with new shavings, or stir your deep-litter bedding and top it with dry, carbon-rich material to maintain a healthy balance.
- **Feed storage check:** Inspect your feed bins. Look for mouse droppings, chew marks on the plastic, or moisture that might have breached the seal and caused mold.
- **Run inspection:** Investigate the coop and run for dampness, early signs of flies, rodent burrows along the fence line, or structural damage to the hardware cloth.
- **Odor, fly, and rodent check:** Stand near the run and observe. Are flies accumulating? Is there a lingering smell? Address moisture and droppings immediately before these escalate into a neighbor complaint.
- **Record review:** Take five minutes to sit with the flock and log any changes in egg production, feed consumption, or behavior.

Monthly and seasonal tasks

Pick one weekend a month for a deeper inspection, general maintenance, and a broader review of your flock's impact.

- **Deeper inspection:** Empty and scrub the feeders and waterers using warm water and a poultry-safe cleaner used according to label directions. Brush down the roosting bars and check for signs of mites or lice in the crevices. Examine hardware, hinges, and locks for rust or looseness.

- **Supplies and repairs:** Check your inventory of feed, bedding, and first-aid supplies. Repair any sagging wire, loose trim, or gaps in the roofing.
- **Neighbor-facing check:** Walk your property line. Look at your chicken setup strictly from your neighbor's perspective. Is the run looking untidy? Are there flies lingering near the shared fence? Address these issues proactively before anyone knocks on your door.

Four times a year, adjust your management for the changing environment:

- **Weather preparation:** Prepare your waterers for freezing winter temperatures, or adjust your shade structures and ventilation for the sweltering summer heat.
- **Storm readiness:** Check your roof panels, secure loose tarps, and clear drainage paths around the run to prevent flooding.
- **Predator pressure review:** In many areas, predator activity changes as food sources and weather shift. Double-check wire integrity and your trenching apron before fall and winter.

Routine design for real life

A routine only works if it survives real life. You need a plan that holds up when the morning rush turns chaotic, when you are sick in bed, or when a massive storm rolls in just as you are supposed to lock up. Design your routine for your worst day, not your ideal weekend. Think through these scenarios now, while the stakes are low: Who handles morning care on work days when everyone is running late? Who handles lockup if you are home late? What happens if the usual caretaker is sick with the flu?

If you have children in the house, backyard chickens offer incredible lessons in responsibility and animal care. But it is a mistake to assume kids can handle the flock on their own. An adult must hold final responsibility for the flock's safety and health. Children lose interest, get distracted, or fail to notice subtle warning signs like a hen walking with a slight limp. Children can absolutely help with tasks, provided their maturity level and the local setup make it safe. However, do not assign risky tasks without direct adult supervision. Manage household boundaries, egg collection, and handwashing rules strictly according to the guidelines in Chapter 20, and ensure an adult is verifying that water is clean, bedding is dry, and doors are locked.

Vacation and sitters

Regarding travel, a common beginner mistake is assuming vacation care is easy to improvise. It is not. You cannot simply ask a neighbor to toss scratch into the yard while you are in another state for a week. Your flock needs someone who knows the daily scan, the lockup routine, and the emergency protocol. When you are out of town, your pet sitter needs clear, written instructions. Do not rely on verbal explanations. Your written sitter checklist must include exact feeding amounts, latch instructions, bird counts, and clear protocols for what to do if a bird is missing, sick, or injured. If a bird appears sick or injured, the sitter should isolate it from the flock if this can be done safely, then call you and the designated poultry or avian veterinarian before moving the bird.

For sudden deaths, multiple sick birds, respiratory or neurologic signs, or suspected reportable disease, the sitter should contact the veterinarian and the state animal-health or sick-bird hotline, or USDA, as directed. Do not move the bird off the property until a veterinarian, clinic, diagnostic lab, or animal-health official gives instructions; if transport is approved, follow their containment and biosecurity directions. For severe injury or distress, the sitter should call the veterinary clinic or emergency service immediately so they can triage and prepare.

Before you travel, have your poultry or avian veterinarian, Extension contact, or local animal-health office review this sick-bird transport plan, because the right action can depend on the severity of the situation, suspected disease, current outbreak or quarantine status, local requirements, and clinic policy.

Chapter 18: Move from first eggs to steady habits in days 31 to 90

You open the nesting box lid, push aside a handful of pine shavings, and see it. Your first egg. It might be smaller than a standard grocery store egg, or its shell might be slightly speckled, but finding it is the moment every suburban chicken keeper anticipates.

But while the first egg is an excellent biological milestone, it is not the finish line.

Many beginners assume that finding an egg means their setup is flawless and the hardest work is officially behind them. You might walk out to the coop, retrieve your first egg, and simultaneously ignore that the bedding near the waterer is soaked, the feeder has been tipped over yet again, and the latch on the run door now requires two hands and a hip bump to force shut.

The true goal of your first ninety days is not extracting a single egg. The goal is establishing a calm, predictable household routine that functions smoothly long after the initial excitement fades. The first egg is just a checkpoint on the way to a permanent operating plan. Steady, manageable habits are what make your suburban flock sustainable.

First eggs without guesswork

Let go of the assumption that egg production will be perfectly steady right out of the gate. For some flocks, laying begins comfortably within this first ninety-day window. For others, it takes much longer. The exact timing of a hen's first egg varies significantly depending on the age of your pullets, the specific breed, the season, available daylight hours, environmental stress, and the individual bird.

When young hens finally do begin laying, their early efforts are notoriously unpredictable. You may find an egg on Tuesday and nothing else until the following week. You might discover tiny "fairy" eggs, misshapen shells, or eggs with soft, rubbery casings instead of hard exterior shells. This early irregularity is a normal part of a pullet's reproductive system maturing and coming online.

Minor fluctuations in size, shape, and frequency are standard during this window. However, overall behavior always matters more than the egg count itself. If irregularities continue for prolonged periods, or if changes in laying are accompanied by lethargy, appetite loss, physical swelling, or visible distress, do not rely on your daily egg tally to gauge the health of your flock. Contact a poultry veterinarian or your local Extension office for guidance.

When laying does stabilize, collect the eggs daily to prevent them from being stepped on, soiled, or eaten by the birds. Collect them with clean hands, set aside any cracked or heavily soiled eggs immediately, and track your general egg count pattern as part of your flock records. We will cover the specific, conservative household defaults for safe egg handling, washing, and refrigeration in Chapter 19.

The day-30 and day-90 reviews

During the first thirty days of chicken keeping, beginners often power through minor inconveniences fueled purely by excitement. If you have to haul water across the yard in a leaky bucket, you do it with a smile. If you have children, they likely fought over who got the privilege of filling the feeder every morning.

By day thirty, this novelty wears off. The children who begged to do chores are suddenly busy when it is time to rake out the run. The leaky bucket is no longer a fun quirk; it is a daily annoyance. This transition is entirely normal. Day thirty is the exact moment to gather your household, update your expectations, and correct the physical setup before temporary annoyances become permanent frustrations.

By day ninety, the honeymoon phase is entirely over. Your birds have established their pecking order, chosen their preferred roosting bars, and decided which corners of the run they like to scratch in the most. The physical footprint of your flock is fully set. To shift from arrival mode to stable operation, you need a systematic way to evaluate your daily effort.

The routine review worksheet tool

The 30-day and 90-day reviews are designed to identify exactly what is breaking down in your routine. You will find the printable version of this worksheet in Appendix G. Use it at the thirty-day mark to conduct a friction audit, and again at the ninety-day mark to finalize your permanent household operating plan. Walk through your setup and answer the following operational questions:

Feed and water stability: Are your feeders and waterers functioning properly? If the birds have tipped the feeder over multiple times, adjust its height or secure it to a fixed post. Assess whether your current water capacity actually lasts a full day without requiring an emergency refill during your working hours. Check the ground beneath the waterer for chronic leaks that ruin bedding.

Bedding and cleaning intervals: The cleaning schedule that felt like a wild guess in week one is now driven by visible facts. You can see exactly how fast manure builds up under the roosting bars. Check the coop corners and the run floor. Are there lingering ammonia smells? Decide whether you need to increase the depth of your pine shavings or adjust the frequency of your cleaning route based on the actual waste load.

Predator resistance check: Walk the perimeter of your run and coop. Look for fresh scratch marks, newly dug holes along the hardware cloth skirt, or signs of chewing near the baseboards. Check every latch, hinge, and fastener. Wood swells and shifts as weather changes; a gate latch that closed perfectly on day one might now be misaligned, leaving a dangerous gap.

Neighbor impact and property lines: Take ten minutes to perform a property-line check from your neighbor's perspective. Walk the shared fence. What do you smell? Can you hear the birds clearly from the property line? Are there more flies present than there were a month ago? How visible is the run, and does it look clean from their angle? Proactively address odor, noise, or visual clutter before a neighbor feels compelled to complain.

Household boundaries: Ensure that children, family dogs, and guests are still respecting the safety and hygiene boundaries established on day one. Familiarity often breeds complacency. Re-establish strict rules regarding handwashing, shoe-changing, and keeping the family dog separated from the run.

Budget reality and finalizing the plan

By the ninety-day mark, you also have enough receipts to run a definitive budget reality check. Pull out the estimates you made before you bought your birds and compare them to your actual spending. Calculate the exact burn rate of your feed and bedding. Note how often you need to purchase fresh pine shavings and log any unexpected repairs or replacement supplies.

Tracking these consumable costs is strictly for long-term supply management. Avoid the mental trap of framing your fresh eggs as financial payback. You are not offsetting the cost of a predator-resistant wooden coop, hardware cloth, and premium feed with a few dozen backyard eggs. Keep your budget grounded in the reality of responsible animal care rather than focusing on a fictional return on investment.

With your ninety-day review complete, turn to Appendix K to lock in these habits. Take the answers from your review worksheet and formally update your household management strategy. Conduct a final legal review to confirm that your actual flock size, coop placement, and waste management plan still comply with your municipal, county, and homeowner's association requirements.

Document the exact daily, weekly, and seasonal maintenance your coop requires based on three months of hard evidence. List your emergency protocols, update the contact information for your trusted veterinary or Extension resources, and finalize the written instructions you will give to a caregiver when you take a vacation.

You have successfully navigated the steepest part of the beginner's learning curve. You survived the research phase, the physical setup, the legal verification, and the arrival of your birds. You have adapted to the daily rhythm of animal care and held your first eggs. Be proud of the daily system you have built, and stay grounded in the reality of what it takes to maintain it safely.

Now that the flock is physically secure and your backyard routine is set, your focus must shift toward food safety. It is time to connect these backyard habits directly to your kitchen. In Part 5, we will move into the practical realities of living well with chickens, starting with the single, conservative household standard for safely collecting, handling, washing, and storing your daily egg harvest.

My Final Notes on Bringing Birds Home

Part 5: Live well with hens in a neighborhood

THE FIRST FEW WEEKS of backyard chicken keeping usually feel like a highly focused sprint. You researched local ordinances, mapped out property lines, built a secure coop, and navigated the specific anxieties of bringing home fragile chicks or young pullets. You spent weeks carefully establishing a daily routine. In those early days, every time you walked out the back door, your primary thought was likely a simple, persistent question: "Are my birds okay?"

Eventually, the startup phase ends. Daily chores become muscle memory, and you stop checking the waterer every two hours. You stop assuming every drop in temperature spells immediate disaster. This marks the crucial moment your backyard flock stops being an all-consuming project and settles into everyday household life.

When the novelty of a new neighborhood coop fades, the reality of daily management remains. A thriving suburban flock should not just survive; it needs to blend smoothly into the background of your street and the rhythm of your home. You are no longer just trying to keep birds alive. You are establishing long-term systems to ensure your neighbors remain unbothered, your yard remains usable, and your household stays healthy.

The neighbor-peace concept in practice

Earlier in this book, we introduced the concept of neighbor peace as a foundational part of your decision to keep hens. When you verified local laws and placed your coop, your neighbors were mostly hypothetical participants in your project. Now, they are living next to the reality of your flock. Maintaining strong community relationships is not a one-time task you checked off during the planning phase; it is an ongoing, daily practice.

Odor, noise, and pests are the quickest ways to erode goodwill with the people living on the other side of your shared fence. If you can smell your coop from your patio, your neighbors can likely smell it from theirs. If your daily routines allow loose feed to attract rodents, those pests will eventually cross property lines. Your primary objective in this final phase is to be known as the thoughtful household that occasionally shares a carton of fresh eggs, not the problematic residence that introduced a localized nuisance to the subdivision.

Proper routines ensure your flock remains a private joy rather than a public frustration. When waste is properly managed, bedding is kept dry, and feed is stored securely, your chickens remain an asset to your property rather than a liability to the neighborhood.

Transitioning to the final operating plan

Up until now, your focus has been primarily external: buying the right equipment, choosing the right breed, and assembling the right physical infrastructure. Now, your focus must shift to the internal operations of your household. You are transitioning from building a setup to running an operating plan.

Many beginners mistakenly believe that once the coop is built and the birds are integrated, the work is simply tossing out feed and collecting eggs. But a suburban yard is a multi-use environment. You share your yard with children, pets, gardens, and guests, as well as local wildlife and seasonal weather patterns. A successful flock owner manages all of these intersections proactively.

Your operating plan is the set of physical boundaries and daily routines that keeps the flock contained and the household secure. It means establishing one conservative household default for managing risk, rather than improvising every time a new challenge arises. If you rely on guessing, you will eventually make a mistake that affects your family's health, your pets' safety, or your neighbors' patience. A written, predictable plan, which you will finalize using the templates in your companion PDF, removes the guesswork from suburban chicken keeping.

The roadmap for neighborhood living

The remaining chapters of this book are designed to guide you through this final transition. They do not introduce new construction projects or new purchasing decisions; instead, they provide the operational boundaries required to live well with hens in a neighborhood context. We will move through five distinct areas of daily and seasonal management.

Managing the harvest and household hygiene: As pullets mature into laying hens, regular egg collection brings the reality of the backyard straight into your kitchen. Chapter 19 establishes exactly what it takes to move safely from the dusty chicken run to the kitchen sink. You will set a single, conservative household standard for handling eggs, washing hands, and keeping outdoor pathogens strictly separated from vulnerable family members and food preparation areas.

Navigating the shared yard: Chickens are curious, relentless foragers who do not understand property lines, prized vegetable beds, or the unpredictability of family pets. Chapter 20 addresses the physical and social boundaries necessary for a multi-use yard. You will learn how to navigate the presence of children, dogs, gardens, and guests, including understanding the limits of dog behavior around prey animals and establishing firm rules for visitors.

Observing the baseline and spotting trouble: Instead of anxiously checking on the birds every hour, your daily observations will evolve into calm awareness. Chapter 21 focuses on learning your flock's normal baseline rhythms so you can instantly spot early warning signs of illness or distress. You will learn how to separate a normal daily scan from actionable warning categories, whether dealing with a single injured bird, an unexplained drop in production, or the sudden onset of multiple sick birds.

Knowing exactly who to call: No matter how well you manage your daily routine, you may eventually encounter a health or safety situation that exceeds your experience. When that happens, crowdsourcing answers on the internet is not a viable strategy. Chapter 22 provides a clear escalation path for seeking professional help. You will learn how to distinguish between issues requiring a poultry veterinarian, a local Extension office, an agricultural diagnostic lab, or animal control. You will also learn the critical protocols for documenting symptoms and knowing when to call before transporting a sick bird or when movement may be inappropriate without veterinary, diagnostic-lab, or animal-health instructions.

Adapting to seasons and absences: A suburban backyard is a dynamic environment, subject to shifting seasons, extreme weather, and the reality of your family's schedule. Chapter 23 covers the necessary adjustments for long-term flock management. You will learn how to safely adapt your setup for heat, cold, and seasonal changes without relying on dangerous equipment or unverified advice. Just as importantly, you will learn how to document your routines so a friend or hired pet sitter can safely manage your flock during family vacations.

The final step

You have already done the heavy lifting. You verified your local laws, mapped your yard, built the infrastructure, and navigated the delicate early weeks. The daily habits are falling into place, and the birds are finally thriving in their environment.

The final step in your journey is simply maintaining these practical systems. When your neighborhood relations are secure, your household boundaries are firm, and your yard is manageable, your flock will bring fresh food and connection to your home without adding chronic stress to your life. The setup is complete. Now, it is time to live well with your flock.

Chapter 19: Handle eggs and hygiene without guesswork

PICTURE WALKING OUT TO your backyard on a crisp morning. You step up to the coop, open the nesting box lid, and pull out a warm, brown egg. This experience feels rewarding and incredibly natural. It is exactly the moment you envisioned when you decided to bring chickens into your life.

But right at this moment, it is easy to fall into a common beginner trap. Because the egg came from your own backyard, and because your hens look vibrant and glossy, you might assume your homegrown harvest is automatically cleaner, healthier, or lower-risk than eggs sitting at the grocery store.

They are not. "From your backyard" does not mean "exempt from food-safety rules."

Handling fresh eggs and maintaining a healthy household requires a deliberate, uncompromising routine. Establishing these habits ensures that keeping chickens remains a safe, low-stress experience for your entire family and anyone who visits your home.

Healthy-looking birds still carry germs

Poultry routinely carry bacteria like *Salmonella* and *Campylobacter* in their digestive tracts, even if the birds look perfectly energetic and healthy [VERIFY: CDC/USDA public health data]. These bacteria are naturally shed in poultry droppings, which means they can easily end up on feathers, feet, coop bedding, and the eggs your hens lay.

A hen can carry these germs and show absolutely no signs of illness. When new chicken keepers hear this, they often assume that a clean coop and healthy-looking birds mean there is zero risk. Public health experts strongly warn against this assumption. The presence of these bacteria is a normal part of poultry biology. It is not a sign that you are a bad chicken keeper or that your flock is diseased.

Think of backyard poultry hygiene just like handling raw chicken in your kitchen or picking up after a family dog. You do not panic when you place raw poultry on a cutting board; you simply wash the board and wash your hands. Ordinary, consistent caution is all it takes to protect everyone.

The physical boundaries

Because suburban chickens can feel like pets, the line between backyard livestock and house animal easily blurs. You might be tempted to bring a docile hen inside the house to show a visitor, or let a newly hatched chick run across the kitchen island for a photo.

Chickens belong outside. You must establish clear, physical boundaries between your flock and your living spaces. Keep all poultry out of your home, and absolutely keep them out of kitchens, dining rooms, and any areas where human food is prepared or consumed.

This strict boundary applies to your equipment, too. Never wash a dirty poultry waterer, feed trough, or egg-collection basket in your kitchen sink. If you need to scrub coop equipment, use an outdoor hose, a dedicated utility sink in the garage, or a wash bucket in the yard.

One of the best habits you can adopt is the "coop shoes" routine. Dedicate a specific pair of boots or slip-on garden shoes entirely to chicken chores. When you walk out to the run, you wear those shoes. When you walk back to the house, those shoes stay outside on the porch or in the mudroom. They never cross the threshold into your living areas. This simple physical barrier prevents you from tracking microscopic poultry waste across your floors and carpets.

Establish a hygiene station

A reliable handwashing routine is your best defense against illness. Every member of the household, and every guest, needs to understand exactly when and how handwashing is required.

Make it a non-negotiable house rule to wash hands thoroughly with soap and warm water immediately after:

- Touching your birds, whether holding a chick or petting an adult hen.
- Touching anything inside the coop or run, including feeders, waterers, latches, and bedding.
- Collecting, washing, or handling unwashed eggs.
- Working in the soil directly adjacent to the chicken run.

To make this seamless, set up a practical handwashing station [ASSET REQUIRED: Hygiene station diagram]. This usually means stocking a dedicated pump of soap and paper towels at the nearest utility sink or mudroom basin. Hand sanitizer can live near the coop gate as a temporary backup, but only use it when soap and water are not immediately accessible, and always follow up with a proper handwash when you go inside.

Protect vulnerable household members

For healthy adults, many poultry-associated infections are mild, but infections such as *Salmonella* or *Campylobacter* can become severe and, in some cases, life-threatening, especially for young children, older adults, pregnant people, and people with weakened immune systems.

You must take strict extra precautions if your household includes or is frequently visited by:

- Pregnant people.
- Children younger than five.
- Adults 65 and older.
- Anyone with a compromised or weakened immune system.

If you have vulnerable people in your home, assign higher-risk chores to a lower-risk adult. As a conservative household rule, vulnerable household members should not clean the coop, handle dirty coop equipment, or process unwashed eggs [VERIFY: CDC high-risk poultry guidance].

Public health guidance also advises against kissing or snuggling poultry near the face. Teach children to interact with the birds by tossing treats into the run or petting them gently on the back, keeping their hands completely away from their faces until they have marched straight to the sink. Furthermore, any eggs consumed by vulnerable individuals must be cooked thoroughly until both the yolk and the white are completely firm. Over-easy eggs or recipes requiring raw egg whites should be avoided, or made using store-bought pasteurized eggs.

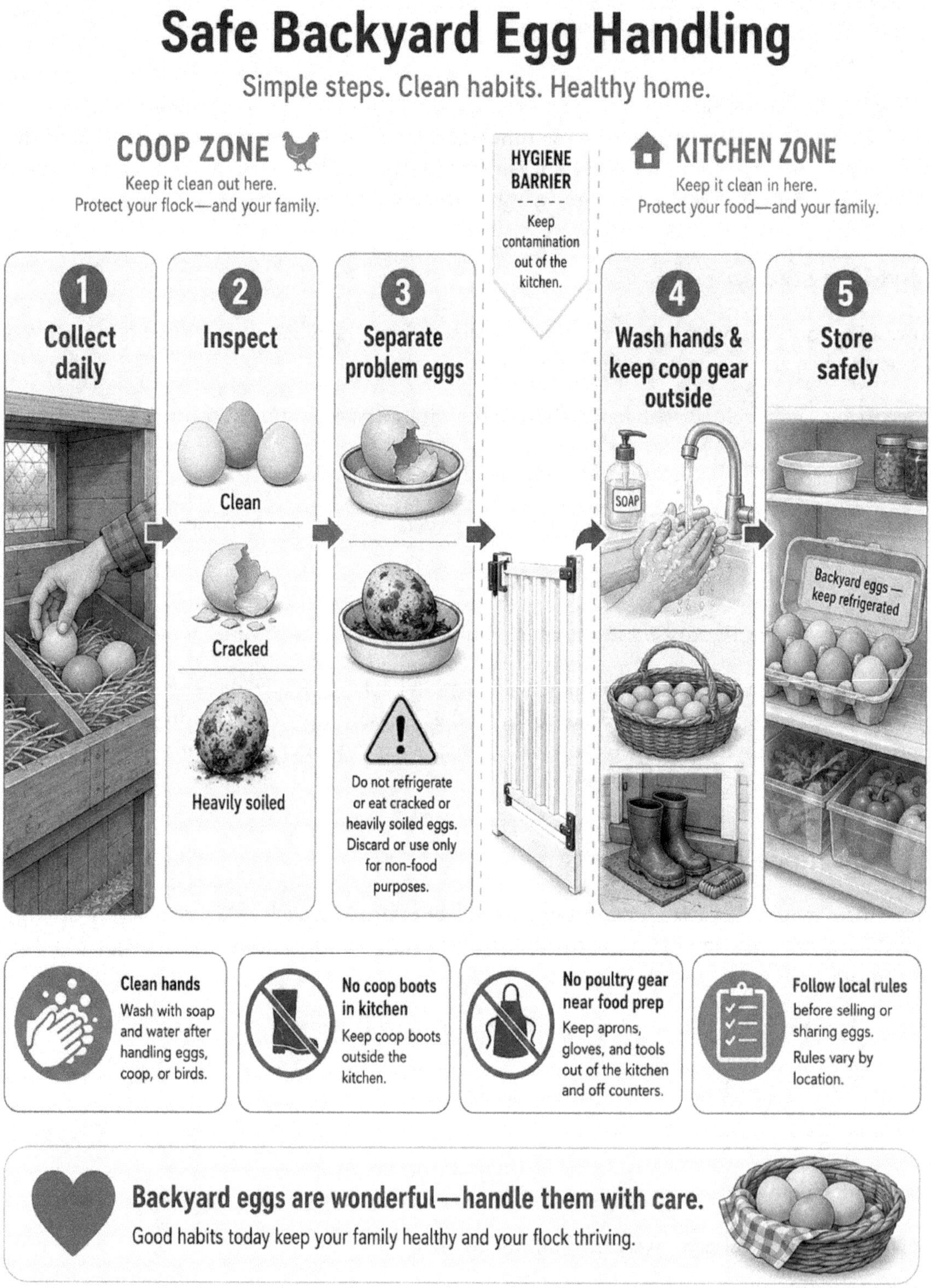

One conservative egg-handling default

If you search the internet for egg-washing advice, you will find endless debates. Some keepers insist on leaving unwashed eggs in beautiful ceramic bowls on the kitchen counter. Others insist on soaking them in special chemical washes.

When a hen lays an egg, her body coats the shell in a wet, invisible layer called the cuticle, or bloom. As this layer dries, it helps seal the microscopic pores on the eggshell, which helps keep bacteria out. Wet washing can remove much of the egg's protective cuticle, or bloom. After wet-cleaning an egg, dry it and refrigerate it promptly at 40°F or below to slow bacterial growth. For a conservative U.S. backyard-flock default, refrigerate collected eggs after cleaning and handling, and follow any state or local egg-sale rules.

Because expert recommendations can vary, treating casual internet advice as a food-safety standard is dangerous. For a suburban backyard flock, the smartest approach is to bypass the guesswork entirely and establish **one conservative household default** aligned with public health best practices. Use the following protocol for every egg, every time [see Appendix M].

Step 1: Collect daily in a clean container. Check your nest boxes at least once a day, though twice is better. Leaving eggs in the coop increases the chance that a hen will step on them with dirty feet or accidentally crack a shell. Use a dedicated, washable wire basket or a plastic container that can be sanitized easily.

Step 2: Inspect and discard the risks. Most days, you will find clean, intact shells. But occasionally, you will find an egg with a hairline crack or a crushed side. Do not attempt to salvage it. Do not feed it to your pets. Bacteria easily pass through compromised shells into the yolk. Discard cracked eggs immediately in the household trash. Similarly, if an egg is heavily caked in mud or entirely covered in feces, discard it. It is not worth the risk of bringing heavily contaminated material into your kitchen.

Step 3: Wash with warm running water. For the intact, normal eggs you bring inside, washing is the safest conservative default. Wash your eggs individually under potable running water that is warmer than the egg. Authoritative guidance commonly specifies at least 90°F and about 10°F to 20°F warmer than the egg; some Extension guidance recommends roughly 100°F to 120°F. Do not wash eggs in water colder than the egg. Cooler water can cause the egg contents to contract, which may draw contaminated water or bacteria through shell pores. Do not soak eggs in standing water.

Step 4: Dry and refrigerate promptly. Once wet-cleaned, dry the eggs completely with clean paper towels. Because washing can remove much of the protective bloom, the eggs should go promptly into the refrigerator at 40°F or below. Store them in a clean, dedicated carton on an inside shelf of the fridge, not in the door where temperatures fluctuate.

By sticking to this single default, collect, sort, wash warm, dry, and refrigerate, you eliminate daily decision fatigue. Everyone in the house knows exactly how an egg is processed, and you maximize the safety of the food you serve your family.

Gifting, sharing, and selling eggs

Eventually, your hens will hit their peak laying stride, and you will likely produce more eggs than your family can reasonably eat. Handing a dozen beautiful, multicolored eggs to a neighbor is a proud moment, but it requires clear communication. When you hand someone a carton, you cannot assume they know the backstory of backyard eggs. Do they know the eggs have been washed? Do they know the protective bloom is gone? Do they know the carton needs to go straight into the fridge?

Take the time to clearly explain how you processed the harvest. Better yet, write a sticky note or use a printed label on the carton that says: *Washed and ready to use. Keep refrigerated.*

If you plan to sell your extra eggs, even just setting up a small cooler at the end of your driveway, be aware that you are stepping into a regulated space. Small-scale agricultural sales are governed by state and local health departments. While many states have exemptions for small backyard flocks, those exemptions still come with non-negotiable rules.

You must proactively verify your state's specific laws regarding egg washing, refrigeration requirements, and labeling. For example, in many jurisdictions, it is strictly illegal to reuse branded grocery-store cartons to sell your backyard eggs, as this violates packaging and traceability laws. You will likely need to purchase fresh, blank cartons and apply a specific label detailing your name, address, and safe-handling instructions.

Finally, never advertise or claim that your backyard eggs are legally safer, cleaner, or healthier than commercial eggs. Stick strictly to the provable facts: your eggs are fresh, they are local, and they came from your backyard flock. When you treat egg handling and household hygiene as a structured, predictable system rather than an afterthought, you protect your family, your neighbors, and your flock's reputation. Once these habits are locked in, you can truly enjoy that morning walk to the coop, knowing exactly what to do with the harvest in your hands.

Chapter 20: Set boundaries for children, dogs, gardens, and guests

A PEACEFUL BACKYARD SHARED by children, pets, gardens, and chickens is a wonderful goal, but "friendly" does not mean "unmanaged." A well-meaning relative might toss inappropriate human food into the run. A child might pet a hen, forget to wash their hands, and grab a snack. A dog's casual trot could escalate into a frantic chase if a bird suddenly flutters its wings. And a curious neighbor could easily release the flock just as the side gate is propped open.

To keep your backyard safe and relaxing, you need firm, clear boundaries. Boundaries are not a sign of distrust; they are how mixed households stay peaceful. They protect your family's health from poultry-borne pathogens. They protect the flock from predators, pets, and stress. And they save your landscaping from being scratched down to bare dirt. When you set limits, you remove the guesswork. Everyone knows where the birds belong, who supervises them, and what areas are off-limits.

Children and public health

Many people start keeping chickens because they want their children to experience nature and learn responsibility. These are excellent goals, but bringing birds and kids together requires an uncompromising hygiene and supervision plan. Live poultry routinely carry bacteria like Salmonella and Campylobacter in their digestive tracts, even if the birds look energetic and perfectly healthy. These bacteria are shed in poultry droppings, which means they can quickly contaminate feathers, feet, eggs, coop bedding, and the surrounding soil.

The Centers for Disease Control and Prevention (CDC) provides clear guidance for handling backyard flocks. Do not let anyone, especially children, kiss or snuggle backyard poultry. Do not hold birds near the face. Children younger than 5, adults 65 and older, pregnant women, and people with weakened immune systems are at higher risk for severe illness from germs associated with live poultry. They should avoid handling or touching chicks, ducklings, adult birds, or items in the birds' living area; if they want to enjoy the flock, have them observe without direct contact and have another healthy adult handle chores.

Handwashing must become a non-negotiable household law. Anytime a child touches a chicken, collects an egg, handles a feeder, or touches the coop enclosure, they must wash their hands with soap and running water before eating, drinking, or touching their face. Hand sanitizer is only a backup for when soap and water are temporarily unavailable, and it does not work effectively if hands are visibly dirty. Keep live poultry and poultry-care equipment, such as feeders, waterers, cages, dirty egg baskets, boots, and coop tools, outside the home and away from kitchens, bathrooms, and food-prep areas. Clean poultry equipment outdoors or in a dedicated utility area. Treat eggshells as potentially contaminated: collect eggs in a dedicated washable container, avoid placing dirty eggs or baskets on food-prep surfaces, wash hands after handling, clean eggs as recommended, and refrigerate eggs promptly.

Do not assume that because your kids are responsible, they can manage the flock alone. Children should never be the sole caretakers of backyard chickens. Age-appropriate chores work best, but only with consistent adult follow-up. Older children can collect eggs or check feed levels, but an adult must verify that the water is clean, the bird count is correct, and the run door is securely latched.

Dogs and other pets

The most common, and most dangerous, objection beginners raise regarding their dogs is: "My dog is gentle."

Individual dog behavior varies wildly, and prey drive is instinctive. A fluttering wing, a sudden squawk, or the erratic running of a frightened hen can trigger an intense, uncontrollable drive in an otherwise affectionate family dog. Even a dog that just wants to play can easily crush a chicken with a single pounce.

Never rely on "he is friendly" as your safety plan. Do not assume that your chickens will eventually learn to avoid the dog. Because individual dog behavior and prey drive vary, you should not assume any dog or any specific training program can make dog-chicken interactions reliably safe [VERIFY: Certified dog behavior professional]. Your management plan should rely on barriers and supervision rather than promises.

No setup removes all risk, but strict physical barriers and supervision are the minimum baseline. Build a dog-boundary plan using these rules:

- No shared unsupervised yard time.
- No free-roaming together in the same space.
- Install secure fencing that the dog cannot push through, jump over, dig under, or paw open.
- Watch for obsessive staring or fence-running along the run. If the dog's presence outside the run causes the flock to panic, adjust the visual boundary by adding a solid barrier so the dog cannot fixate on the birds.

Consider the household that spends weeks letting their calm, senior dog observe the new chickens through a sturdy wire run. Even after the dog shows zero interest, the physical barrier remains in place. They never allow the dog and the chickens to mingle freely. That is what a successful boundary looks like.

Visiting pets are another hazard. If a friend brings their dog over for an afternoon, that dog does not belong in the backyard with your chickens. Keep them entirely separated behind closed doors or gates.

Gardens, landscaping, and yard traffic

Beginners often imagine chickens gently plucking pests from blooming flower beds. In reality, chickens naturally scratch, dig, and dust bathe. Given access to soft, freshly tilled soil, they will carve out deep bowls, uproot seedlings, shred delicate landscaping, and peck at ripening vegetables.

Decide where chickens are allowed before the damage happens. If you want a pristine garden, confine your chickens to a spacious, enclosed run. You can use light garden fencing or standard chicken wire to protect vegetable beds from foraging hens. However, clearly distinguish between garden boundaries and predator protection. Chicken wire is only designed to keep chickens contained; it is not a predator-resistant material and will not stop a dog, raccoon, or fox. Only heavy-gauge hardware cloth attached securely with screws and washers can provide predator resistance for the coop and run.

Managing the yard also means managing waste and chemicals safely. Fresh chicken manure is too high in nitrogen (too "hot") for direct garden application and carries potential pathogens. Compost it securely in a dedicated bin, far from child play areas or pet access, and follow local Extension guidelines for curing times before applying it to edible crops [VERIFY: Extension/IPM reviewer]. Additionally, remove household food scraps from the run before they attract rodents.

If your birds have access to the yard, audit your use of fertilizers, herbicides, and pesticides. Some lawn and garden products, including certain pesticides, herbicides, baits, weed-and-feed products, and treated granules, can harm chickens or create residue risks if birds forage in treated areas. Read and follow the label every time, keep chickens out for the label-specified reentry or grazing interval, and if no interval is listed, keep birds away at least until sprays have dried or dusts and granules have settled. Do not assume a product is safe for poultry unless the label permits that use or an Extension specialist or poultry veterinarian confirms it.

Finally, evaluate your household traffic. Chicken chores happen in the mud, snow, and dark. Map a clear, well-draining path from your back door to the coop. Eliminate trip hazards like exposed outdoor extension cords, scattered tools, or uneven pavers. Route outdoor cords securely away from foot paths, and ensure all outdoor electrical elements utilize GFCI outlets with weather-proof covers.

Guests and visitors

When you have backyard chickens, visitors will naturally be curious. Friends, neighbors, and extended family will want to see the coop, but they rarely know the unwritten rules of suburban chicken keeping.

Without clear instructions, mistakes happen. A guest might open the run door "just for a peek," allowing a hen to slip out just as the neighborhood dog walks by. Set the rules before your guests even step onto the patio.

Visiting children require the exact same adult supervision and strict handwashing rules as your own family. To prevent cross-contamination, politely ask guests to wipe or remove their shoes before coming back into your house, ensuring backyard soil and unseen droppings do not end up on your living room rug.

Build boundary systems into your physical environment. Install self-closing or spring hinges on gates so doors shut automatically if a guest accidentally lets go. Use secure two-step latches, or spring latches secured with carabiners, that a curious toddler or a clever dog cannot quickly pull open.

To help communicate these expectations smoothly, keep a simple guest script ready. As you walk out to the backyard, you can simply say, “We would love to show you the chickens. Just a quick heads-up: please don't stick your fingers through the wire or feed them, and we will all wash our hands with soap and water as soon as we head back inside.”

You can also post a friendly, easy-to-read guest rules card near the back door or directly on the run gate.

Guest Rules for the Coop:

- Please do not open the coop or run; ask the chicken keeper to open them.
- Please do not feed the hens; we will provide approved treats.
- Please do not pick up, touch, or chase the birds.
- Children must stay with an adult at all times.
- Everyone must wash their hands with soap and water after visiting.

Implementing your boundaries

Take a moment to walk your yard and reflect on your daily habits to identify your weak spots. Ask yourself which gates a guest could accidentally leave open, potentially allowing a dog and the chickens to cross paths. Consider which chicken chores are only appropriate for a child if an adult provides follow-up verification. Check your hardware to see where a visiting child might be able to reach a latch.

Use the tools provided in Appendix M: Egg and household hygiene to formalize your setup. By completing the child, dog, and garden boundary plan before the birds arrive, you eliminate the daily anxiety of wondering who left a gate open or what the dog is doing.

Clear boundaries are an act of kindness. They create a calm, predictable environment with less confusion, fewer accidental access points, and clear expectations, allowing everyone to enjoy the backyard safely and in peace.

Chapter 21: Notice normal behavior and early warning signs

When you bring chickens into your backyard, you might assume keeping them safe requires a fully stocked medicine cabinet and a working knowledge of poultry pathology. You might feel pressure to memorize a textbook of avian diseases just in case something goes wrong.

But your most effective tool for maintaining a healthy flock is not a medication or an encyclopedia of symptoms. Your best tool is simply knowing what your birds look like when they are healthy.

Think about how you observe a family dog. If your dog normally greets you at the door with a wagging tail but one evening stays curled up in his bed and refuses dinner, you do not need a veterinary degree to know something is wrong. You instantly recognize a deviation from his baseline behavior.

The same principle applies to backyard hens. You do not need to learn how to diagnose every chicken disease. You only need to build a baseline of normal behavior, notice changes early, categorize the urgency, and know when to ask for qualified help.

Build your normal baseline

To spot a sick bird, you must first study a healthy one. Take time during your first few weeks of keeping chickens to watch your flock without an agenda. You are gathering data on what your specific birds look and sound like on an average day. Focus on these baseline categories:

Eating and drinking: Notice how enthusiastically they approach the feeder in the morning and how often they drink. A healthy flock actively seeks out food and visits the waterer throughout the day.

Movement and posture: A healthy chicken generally stands upright with her tail held high or level, depending on the breed. Her movements look intentional. She explores, scratches the ground, and walks without a limp.

Droppings: Chicken droppings vary widely depending on the weather and diet. Sometimes they are firm with a white cap of urates; other times, they are loose, brown, and foul-smelling after the bird empties her ceca. Learn your flock's normal spectrum so you do not panic over a temporary dietary change.

Feathers: Outside of an annual molt or mating wear, healthy feathers generally look well-kept and lay relatively flat against the body, though some environmental wear is natural.

Social behavior: Learn the normal pecking order. Know who eats first, who chases whom, and who prefers to roost on the highest bar. Healthy hens are social and stay relatively close to the flock.

Laying changes: A healthy hen might pause her laying because of extreme heat, decreasing daylight, broodiness, or a molt. Do not assume a drop in egg production automatically points to illness, but log it as part of your baseline data. Laying changes matter most when paired with other signs like lethargy.

Vocalization: Your yard has a baseline soundtrack of soft clucks, occasional purrs, and the loud "egg song" they sing after laying. Familiarize yourself with these normal sounds so you can easily distinguish them from distressed squawks, wheezing, or unusual silence.

Weather-related behavior: Know how your birds react to the elements. In extreme heat, it is normal to see them panting with their beaks open and wings held slightly away from their bodies. In cold weather, they may fluff their feathers to trap warm air and tuck one foot into their breast feathers.

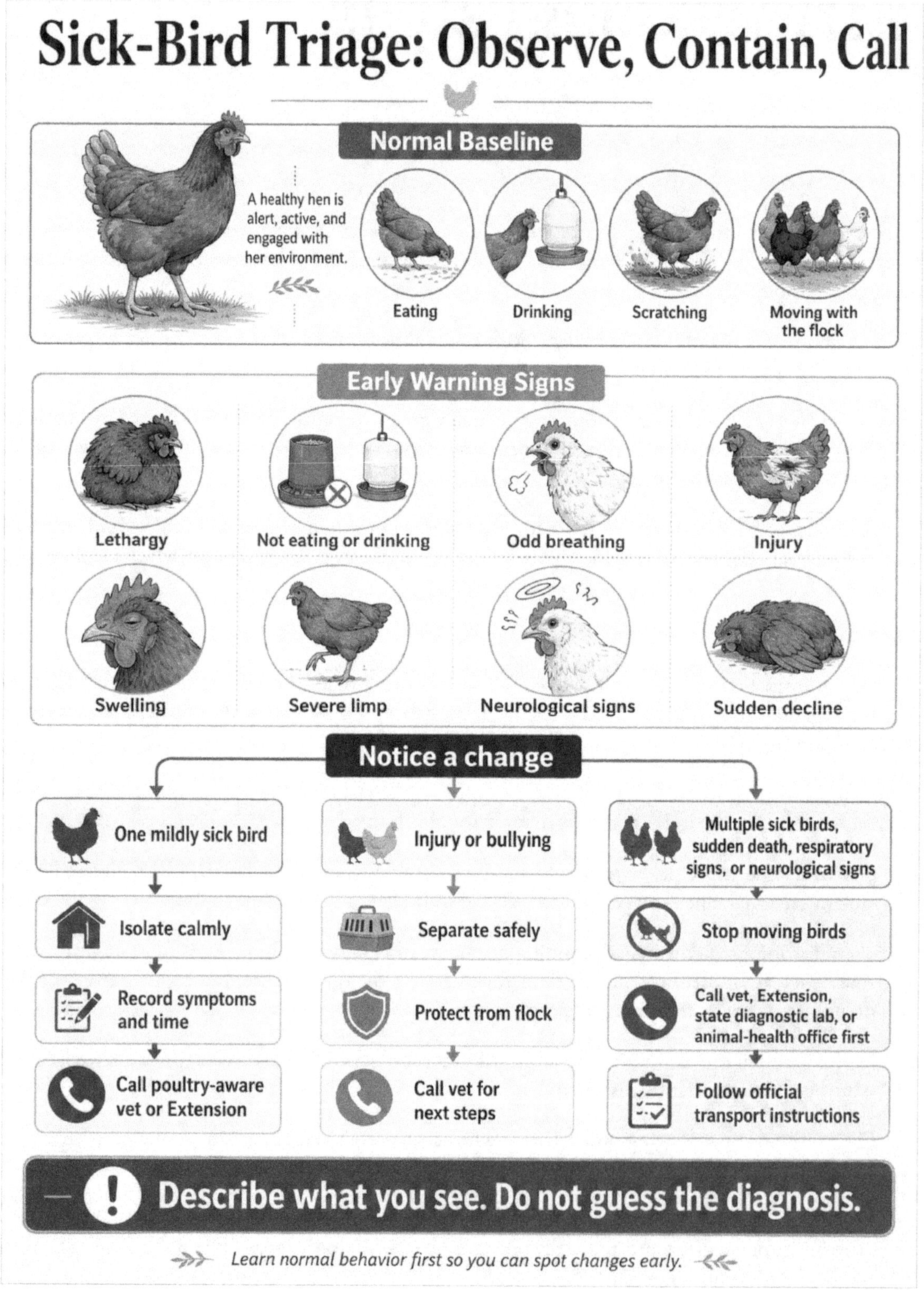

The daily 60-second scan

Once you know what normal looks like, you can build a daily observation habit. Stand outside the run or coop for one minute before you open the door or rattle the feed scoop. Watch them before you interrupt them.

During this short daily routine, use a 60-second scan:

- Who comes to the food and water?
- Who hangs back?
- Is anyone breathing oddly or making a rattling sound?
- Is anyone limping, holding a wing awkwardly, or standing puffed up in a corner?
- Are the visible droppings broadly normal?
- Did anything change from yesterday?

Ask yourself two questions as you watch: What does normal morning behavior look like for your flock? And which bird would you notice first if she hung back?

By doing this every day, you train your brain to catch subtle shifts long before a bird shows severe distress. If you spot a minor change, make a mental note, write it down in your observation log, and check on that specific bird again in a few hours.

Early warning categories

When your daily observation reveals a change, you need to recognize the warning signs. You are not trying to match these signs to a specific disease; you are simply observing symptoms.

Lethargy: A sick hen will often stand apart from the flock, puff up her feathers, tuck her head under her wing, and close her eyes during active daylight hours.

Not eating or drinking: Chickens have fast metabolisms. Refusing food or water is frequently the first visible sign of distress.

Respiratory signs: If a hen stretches her neck to breathe, bobs her tail with every breath, sneezes persistently, has bubbly eyes, or makes a rattling sound, she needs help.

Injury: Cuts, blood, missing patches of feathers from an attack, or signs of bullying require immediate intervention.

Abnormal swelling: Look for sudden or gradual swelling around the eyes, beak, abdomen, or feet, as well as any unusual discharge from the eyes or nostrils.

Severe lameness: An inability to stand, a sudden severe limp, or an unwillingness to bear weight on one leg is a major warning sign.

Neurological signs: Tremors, walking backward, stargazing (twisting the head back to look straight up), or an inability to balance.

Sudden drop in condition: A bird that rapidly loses weight, drops her comb over, or completely stops interacting with the flock.

Sick-bird triage and containment

How you respond to an early warning sign depends entirely on the scenario. Follow these triage protocols to protect the bird and the rest of the flock.

One mildly sick bird: If one hen stays under the roosting bars, puffed up, and ignores food, do not guess the cause. Isolate her to reduce stress. A clean wire dog crate in a protected outbuilding, garage, or washable area away from pets and children works well. Provide her with her own shallow dishes of fresh water and standard feed. Isolating her protects her from being pecked, allows you to track exactly what she consumes and eliminates, and minimizes exposure to the rest of the flock. Once she is contained, contact a poultry veterinarian or your local Extension office for guidance.

Injury or bullying: If a bird has a visible wound or is actively being bullied, immediately separate her. Chickens are attracted to the color red and will relentlessly peck at a flock mate's wound, making the injury significantly worse. Isolate the injured bird in a safe, warm, and quiet space. Clean the wound as directed by a veterinary professional and do not return her to the flock until she is fully healed and the wound is completely closed and invisible.

Multiple sick birds: If two or more birds show warning signs simultaneously, especially respiratory or neurological signs, elevate your concern immediately. Widespread illness can indicate a contagious issue. If multiple birds are sick, deaths are unusual, or signs suggest HPAI, vND, or another reportable disease, isolate the flock and call a poultry veterinarian, your state veterinarian/state animal health agency, Extension, or USDA before moving any birds. Do not move sick or dead birds off-site unless a veterinarian or animal-health authority instructs you to do so. Moving contagious birds, or contaminated shoes, crates, hands, clothing, tires, or equipment, can spread disease. Practice strict biosecurity, wash your hands, change your shoes, and contact your state veterinarian or diagnostic laboratory immediately.

Sudden death: A sudden unexplained death, and especially multiple deaths over a short period, should be treated as a potential reportable disease or biosecurity concern: isolate the area, use PPE when handling birds or litter, and call a poultry veterinarian, Extension, or your state animal health agency for instructions. Do not handle the deceased bird without wearing disposable gloves and a mask. Double-bag the carcass, store it in a cool, secure place away from scavengers, and wash your hands thoroughly while you wait for instructions. Do not move the carcass off-site unless a veterinarian, diagnostic lab, or animal-health authority tells you how to do it.

Suspected reportable disease: In the United States, HPAI and virulent Newcastle disease are nationally reportable/notifiable poultry diseases. Suspected or confirmed cases should be reported through a veterinarian, diagnostic laboratory, state animal health agency/state veterinarian, or USDA, and state or federal animal-health officials direct testing and response. If your flock has sudden high mortality, severe respiratory distress, purple or bluish discoloration of the comb, wattles, or legs, or neurologic signs such as tremors, twisted neck, incoordination, stumbling, falling, or paralysis, call your state veterinarian/state animal health agency or USDA's sick-bird/FAD hotline at 1-866-536-7593 immediately. Do not move live birds or carcasses off the property unless instructed by animal-health officials or a qualified veterinarian.

Medication and home-remedy cautions

Beginners faced with a sick bird sometimes make mistakes driven by anxiety.

Do not try to diagnose your chicken from internet photos. Describing symptoms on a backyard poultry social media group will yield dozens of conflicting diagnoses. Many diseases look identical to the untrained eye; respiratory issues alone can stem from viruses, bacteria, fungi, or environmental dust.

Do not dose medications casually. Treating a bird with leftover antibiotics, dewormers, or random farm-store medications without a proper diagnosis can mask the real problem, contribute to antibiotic resistance, and compromise food safety.

Whenever a laying hen receives a medication, dewormer, topical chemical, or other treatment, even something bought over the counter at a farm store, assume that egg-withdrawal guidance may apply. An egg-withdrawal period is the interval during which eggs from a treated hen should be discarded or withheld from use because residues may be present in eggs. Do not consume, share, or sell eggs from a treated bird until you have confirmed the correct withdrawal guidance from the product label and, when in doubt, a poultry-savvy veterinarian.

Furthermore, do not rely on home remedies as a substitute for qualified help. Adding garlic, herbs, essential oils, or internet tonics to the feed or water will not cure a severe infection, fix a broken bone, or resolve a bound egg. "Natural" does not mean "medically effective." If a bird is genuinely sick, she needs an accurate diagnosis and appropriate care, not a pantry experiment.

The sick-bird observation log and tools

When you isolate a bird and prepare to call for advice, you need organized information. In Appendix O you will find a Sick-Bird Triage Table and a Baseline Observation Log.

Before you call a veterinarian or Extension agent, write down the bird's identification and description. Note the exact date and time you first noticed the issue. Record every sign you observed, detailing her appetite, water intake, and droppings. Crucially, note any recent environmental changes: a new bag of feed, variations in water access, extreme heat or cold, ventilation issues, a predator scare, newly introduced birds, cleaning products, or a recent coop move.

Having this information written down keeps you focused. You do not have to rely on memory to recall whether the hen sneezed on Tuesday or Wednesday. You hand over a clear, organized baseline of facts.

Your job as a backyard flock keeper is not to name the disease or dose the cure. Your job is to observe your birds daily, notice changes early, protect the vulnerable, record what you see, and ask for the right help.

Early Notes on My Birds

Chapter 22: Know when to call for help and what information to bring

Imagine you notice one of your hens sitting quietly in the corner of the run. Her feathers are puffed out, she is ignoring the treats you scatter, and her breathing seems slightly labored. Your first instinct might be to pull out your phone, type a quick search, and brace yourself.

Within minutes, you are scrolling through a forum filled with conflicting, urgent advice. One commenter tells you to immediately dose the entire flock with a leftover livestock antibiotic. Another insists you add garlic to their water. A third confidently declares it to be a highly contagious virus and tells you the flock is doomed. Your panic rises.

Resist the urge to fall down an internet rabbit hole. Online communities are excellent for coop-building inspiration and moral support, but they cannot reliably diagnose an infection from a paragraph of text. Instead of guessing, your job is to use the baseline data you gathered during your daily observations and contact a qualified professional. You do not need to be a poultry pathologist to keep chickens safely, but you do need a help-seeking plan in place before an emergency strikes.

Before you even pick up the phone, understand this critical distinction: routine, isolated injuries or mild illness may still require veterinary care, but multiple sick birds, unusual deaths, or signs suggestive of reportable disease require a call before any transport. For multiple sick birds, unusual deaths, or signs suggestive of reportable disease, call a poultry veterinarian, diagnostic lab, state animal health agency, or USDA first, and do not transport live birds or carcasses unless they instruct you to do so. For a routine, isolated injury or illness, call the veterinarian before arriving so the clinic can give biosecurity and intake instructions.

When a family dog is sick, the default action is to put him in the car and drive to the veterinary clinic. With poultry, doing this in the wrong situation can trigger a biosecurity disaster. If your flock is experiencing sudden death, if multiple birds fall ill at the same time, or if you suspect a highly contagious respiratory disease, putting a sick or dead bird into a carrier and driving across town is dangerous. You could inadvertently spread pathogens, such as Highly Pathogenic Avian Influenza (HPAI) or virulent Newcastle disease, on birds, carcasses, crates, shoes, clothing, hands, tires, or equipment to other local flocks, agricultural supply stores, or a veterinary waiting room.

Until you receive instructions, keep the sick bird isolated at home, secure the rest of the flock, wash your hands thoroughly, and make a phone call first. A professional will tell you whether it is safe to transport the bird, whether they need to come to you, or whether you must follow specific safe-handling protocols for submitting a sample to a diagnostic lab. Use current official biosecurity guidance from USDA APHIS and your state animal health agency or department of agriculture, and follow the instructions of the professionals handling your case because contacts and rules can change.

Beginners often assume that without a dedicated poultry veterinarian in town, they are completely on their own. In reality, you have a wide network of qualified help available, provided you understand who handles what.

- **Poultry or General Veterinarian:** Your veterinarian is the primary contact for individual bird trauma, severe injuries, or targeted medical treatments. Finding a board-certified avian vet can be challenging, but call local dog-and-cat clinics; they may have a staff member who treats backyard poultry or can provide a referral. Do not assume immediate availability; establish this relationship before a bird gets sick.

- **University Extension Office:** Your local agricultural Extension specialists are invaluable for general flock management, husbandry education, and regional biosecurity updates. They cannot diagnose diseases over the phone or prescribe medications, but they are often the best starting point for finding local testing facilities.

- **State Diagnostic Laboratory:** When you need formal testing or a necropsy (an animal autopsy) to determine the exact cause of a bird's death, turn to state or university diagnostic labs. These facilities provide definitive scientific answers. You must call them or check their website to verify submission rules, required packaging, and fees before sending samples.

- **State Veterinarian and Animal Health Agencies:** These officials manage diseases that threaten local agriculture and public health. For suspected reportable poultry disease, avian influenza signs, or sudden unexplained mass mortality, contact your state veterinarian/state animal health agency, a poultry veterinarian, or USDA's sick-bird/FAD hotline at 1-866-536-7593 and follow their instructions. State reporting rules and preferred contacts vary.

- **Animal Control or Municipal Code Enforcement:** Turn to local agencies for legal and community issues. If you receive a noise complaint, need to report a loose neighborhood dog harassing your flock, or have a zoning question, this is your contact. They do not provide veterinary advice.

- **Public Health Department:** If anyone in your household develops an illness after handling the flock or the coop, contact a human health-care provider or your local public-health department immediately. Do not rely on poultry forums for human medical advice regarding potential infections.

- **Emergency Services (911):** Reserve emergency services strictly for immediate human safety threats, such as a major coop fire or severe human injury.

When you call a veterinarian, an Extension agent, or a diagnostic lab, the quality of the help you receive depends on the quality of the information you provide. Think of it like taking a child to urgent care. You do not just walk in and announce a diagnosis; you provide a timeline and a list of symptoms.

Prepare a factual checklist before you dial. Gather the following details:

- **Flock demographics:** The total number of birds you own, their exact ages, and precisely where you acquired them (e.g., hatchery, farm store, neighbor).

- **The timeline:** The exact date and time you first noticed abnormal behavior.

- **Affected numbers:** How many birds are currently showing signs, and how many, if any, have died.

- **Plain descriptions:** Describe what you see, hear, or smell without naming a disease. Say, "Her crop feels like a squishy water balloon, her breath smells fermented, and she has not eaten since yesterday morning," rather than, "I think she has sour crop." Plain descriptions allow the expert to diagnose.

- **Recent changes:** Note any new batches of feed, changes to the water source, extreme weather events, or modifications to the housing.

- **Exposure risks:** Mention recently added birds, wild bird interactions, or visitors to your yard who own poultry.

- **Egg status:** Note whether the hens are still laying, if production has dropped, or if the eggs have changed in shell quality or shape.

- **Interventions:** Honestly list any medications, supplements, or remedies you have already tried.

- **Media:** Have clear photos or short videos ready to send if requested. A short clip of a bird's posture or breathing provides vital context.

When you are on the phone, keep a pen and your Help-call worksheet handy.

Write down the professional's instructions exactly as given. Ask clarifying questions to ensure you understand the next steps: "Should I isolate this bird?" "Is there an egg-withdrawal period for this medication?" "Do I need to submit a sample for testing?"

If the expert directs you to review current biosecurity protocols for avian influenza, ask for the exact website address, usually a USDA APHIS or state agriculture portal, to ensure you are reading official, current guidance rather than outdated blogs. Record the name of the person you spoke with and the date.

Find contact information before you ever need it. Spend fifteen minutes building a concrete plan using your Help-call worksheet. Look up the phone numbers for your closest avian-friendly veterinarian, your county Extension office, the state diagnostic lab, and your local animal control. Keep this list with your flock records or posted near your back door. When a crisis occurs at dusk, you will not have to frantically search the internet for a phone number while holding a sick bird.

Making a clear, fact-based call protects your flock from the dangers of amateur treatment. By understanding who to contact and having your information organized beforehand, you remove the panic from backyard emergencies. Once this safety net is established, you can confidently navigate the predictable, seasonal changes your flock will face.

Do I Know Who to Call and When

Chapter 23: Prepare for weather, vacations, molts, and the next season

AFTER YOUR FIRST NINETY days with backyard chickens, the initial anxiety of setting up the coop fades into a comfortable, predictable rhythm. You know exactly how much feed to scoop, how to handle the morning routine, and what a content flock sounds like. The daily checklist you followed in your operating plan has become second nature.

Then, you flip the calendar.

The evening news forecasts the first severe winter storm. The peak summer sun bakes your backyard. You receive an invitation to a family wedding scheduled three states away. Or you walk out to the coop on a crisp autumn morning and find the run covered in discarded feathers, looking as though a pillow exploded overnight, with every single nesting box empty.

Beginners often assume that once the birds are settled, the heavy lifting is permanently finished. However, a living flock requires ongoing adaptation. Keeping chickens safely requires you to plan for heat, cold, seasonal storms, annual molts, and human travel before they disrupt your routine.

Heat, cold, and weather planning

It is tempting to wait until the forecast drastically changes to prepare for severe temperatures, but early planning prevents panic. Weather strategies must be tailored to your specific region, a suburban backyard in southern Florida requires a completely different approach than a property in upstate New York. Avoid searching for universal temperature rules online. Instead, lean on your local university Extension office, which can provide region-specific guidance on safe wintering and summer heat management.

High heat can become an immediate danger to chickens, sometimes faster than cold in warm climates or poorly shaded yards. Because chickens cannot sweat, they rely on panting and holding their wings away from their bodies to cool down. When the summer sun beats down on your yard, your hens need deep shade, excellent coop ventilation, and constant access to fresh, cool water. Place waterers in fully shaded areas to prevent the water from becoming too hot to drink. You may need to add temporary shade sails over the bare sections of the run or place extra water stations in their favorite resting spots under trees.

When winter arrives, your priorities shift to preventing their drinking water from freezing and blocking direct, freezing wind from blasting through the coop. A common and dangerous beginner mistake is sealing the coop completely tight to trap body heat. A tightly sealed coop traps the moisture generated by the chickens' breath and droppings. That trapped moisture, rather than just the cold air itself, can greatly increase the risk of frostbite on combs and wattles. A good winter-ventilation plan generally allows damp air to escape high in the coop while avoiding cold drafts blowing directly across the birds as they sleep on their roosts; ask your local Extension office or poultry professional for region-specific ventilation guidance.

Electrical and water safety

In freezing climates, keeping water in a liquid state is a daily challenge. Many flock keepers rely on heated water bases or submersible poultry water heaters. While these tools are incredibly useful, they introduce significant electrical and fire risks to a wooden coop filled with dry, combustible bedding.

If you use heated waterers or any outdoor electrical devices, treat them as significant safety equipment, not casual accessories. Use equipment specifically manufactured and rated for outdoor or agricultural use. Follow the manufacturer's instructions exactly. If an outlet supplies power to the coop, have a licensed electrician ensure it is properly installed, weather protected, and GFCI-protected where required for outdoor or wet locations to reduce the risk of severe or fatal shock.

Do not rely on a cheap, indoor-rated extension cord stretched across the lawn to power the coop. Use only cords and devices rated for outdoor use, sized for the electrical load, and routed or protected so they are not exposed to standing water, weather damage, lawnmowers, tripping feet, or chewing rodents. Ensure all connections are shielded from rain and snow. Do not attempt makeshift, do-it-yourself wiring to run electricity to your coop; if you want permanent power, hire a licensed electrician.

Storms and supply reserves

You need a practical plan for heavy rain, snow, and severe storms. A quick physical inspection should become second nature before a major weather event. Walk the perimeter of the coop to ensure latches are secure and the roof is tied down. Check that your drainage channels are clear before heavy rains or snowmelt; proper drainage will save you from hauling heavy, contaminated mud out of the run the next morning.

Furthermore, severe weather can disrupt supply chains or trap you at home. Do not wait until your feed bin is completely empty to buy the next bag. Maintain a reasonable supply reserve so you are never caught without feed, fresh bedding, or a backup waterer during a multi-day storm. Use a [ASSET REQUIRED: Supply reserve checklist] to inventory your essentials before storm season begins.

Molts and seasonal laying variation

If you do not know what to expect, a hen's annual molt can be alarming. Usually occurring in the late summer or autumn as daylight hours decrease, a molt is a natural biological process where chickens shed their old, worn feathers and grow a fresh set for the coming winter.

During this time, your once-glossy hens might look ragged, balding, and exhausted. Growing a new coat of feathers requires immense physical energy and protein. As a result, egg production will drop sharply or stop completely. It is incredibly common for beginners to walk out to an empty nesting box, see a patchy-looking hen, and immediately panic.

Handle a molt calmly. Do not attempt to diagnose an illness based solely on a seasonal drop in eggs. Provide a quiet, low-stress environment. Because feather regrowth requires heavy nutritional support, your local Extension office will likely recommend temporarily switching to a higher-protein feed or providing specific, carefully measured protein supplements during the molt [VERIFY: Extension molt nutrition recommendations].

Remember that egg production naturally fluctuates based on age, season, available daylight, weather, and the physical stress of the molt. If you notice concerning signs that go beyond normal feather loss, such as severe lethargy, respiratory distress, or a total refusal to eat, consult a poultry-savvy veterinarian or your Extension office.

Vacations and backup care

Eventually, you will want to leave your house overnight. A major misconception among beginners is thinking a neighbor can casually toss a handful of kitchen scraps over the fence while you are on a week-long vacation.

Chickens require daily management, even when you travel. Before you book a trip, identify a reliable caretaker to handle the routines that keep your birds safe, contained, and fed.

Do not rely on quick verbal instructions. Provide a highly detailed, written document. This sheet must detail exactly where the supply reserves are stored, how much feed to scoop, and how to safely refill and plug in the waterers. It must explicitly state your evening lockup procedure, including what time the door must be secured and how to visually count and confirm every single bird is inside the hardware-cloth enclosure before walking away. Include your household's specific, conservative default rules for collecting and washing eggs. Provide emergency contact numbers, including your local veterinarian, and clearly define what counts as a true emergency.

Just as importantly, tell your caretaker exactly what "not" to do. Explain that they absolutely must not let the birds free-range while you are away. A loose chicken entering a neighbor's yard when the owner is out of town is a major nuisance liability. Instruct them not to bring their dogs into your yard, not to feed the hens random kitchen scraps, and not to attempt home remedies if a bird looks sick. A strict, boring routine keeps the flock legally contained and physically safe until you return.

The next season and neighborhood review

As you close out one season and prepare for the next, step back from the daily chores and perform a seasonal review. This is your opportunity to formally assess your setup before minor frustrations turn into major, expensive problems.

Inspect your coop and run durability. Check if hinges are rusting, if hardware cloth needs re-securing, or if the roof leaks. Ensure field mice have not compromised your feed storage bins in their search for winter warmth.

Crucially, conduct a seasonal neighbor review. Walk your property line and honestly evaluate how your setup looks and smells from the other side of the fence. In autumn, falling leaves can pile up against your run wire, creating a damp, rotting mass that generates odor and attracts pests. When winter strips the leaves from the trees, your coop might suddenly be highly visible to a neighbor's back patio. Spring rains can push mud toward their property. Keeping the exterior boundary clean and unobtrusive prevents an eyesore from sparking a neighborhood complaint.

Finally, look back at the operating plan you created during your first ninety days. Are you still following it? Maintaining a legal, safe, clean, contained, and neighbor-friendly suburban flock requires ongoing honesty. By planning for weather, establishing vacation protocols, and respecting the natural rhythms of the birds, you stop reacting to surprises. Season by season, managing the flock simply becomes a familiar, seamlessly integrated part of your household life.

Action steps for seasonal success

Use the companion tools to build your seasonal readiness plan before the weather turns:

- **Audit your weather readiness:** Identify your next seasonal risk, heat, cold, storms, or frozen water, and prepare your yard this weekend.
- **Stock your essentials:** Ensure you have enough feed and bedding to survive a multi-day storm or power outage.
- **Prepare your sitter:** Ask yourself: "Who can confidently lock up my hens with 24 hours' notice?" If you do not have an answer, start making calls today.
- **Evaluate your impact:** Walk your property line, inspect your hardware, and ensure you remain compliant with local laws and neighbor expectations.

Conclusion: Your suburban chicken peace plan

In the introduction to this book, you looked out at your lawn and saw a mix of exciting possibilities and potential pitfalls. You saw property lines, curious dogs, close neighbors, and a blank slate. Now, your perspective has entirely shifted. You are no longer relying on hope, social media aesthetics, or guesswork to figure out if backyard chickens are right for your current season of life. You understand exactly how to find out where a coop can legally sit. You have learned how to evaluate the footprint of a secure run, navigate your local municipal ordinances, and lay the groundwork to keep the peace with the people living on the other side of your fence.

You arrived at this book cautiously excited and looking for clarity. You are leaving equipped to make a grounded, factual decision. Reaching the final pages does not mean you are automatically ready to drive to the farm store or order a box of day-old chicks. Taking your time now is not a sign of hesitation; it is the mark of a responsible, prepared flock owner. Good chicken keeping always begins long before the first birds come home. It is planned, verified, budgeted, and maintained. Rather than guessing your way through the process, you now understand how to decide if a flock fits your yard, your wallet, and your daily schedule.

Instead of relying on neighborhood rumors or outdated internet advice, you know how to track down the actual rules for your specific address. You can make a neighbor-friendly promise that you can actually keep, focusing relentlessly on odor control, secure containment, pest management, and minimizing noise. You have learned to avoid the allure of "chicken math" by selecting a flock size and breed type that fits your specific climate, rather than maxing out your legal limits by default. Most importantly, you know the undeniable value of building the daily, weekly, and monthly routines that prevent health and safety problems before they start.

Before welcoming a single bird onto your property, you need a finalized operating plan. This is a written document compiling your legal status, yard placement, flock size, supplies, routines, cleaning methods, budget, and emergency contacts. In Appendix K, you will find a completed sample first-flock operating plan to use as a model. You should complete your own operating plan before purchasing any live animals. If any item on your checklist is blank, uncertain, or based on a hopeful guess, pause. Do not buy birds yet.

To help you finalize that document and arrive at your ultimate decision, run through this comprehensive readiness check:

First: The legal record. You have documented your municipal, county, zoning, and homeowner association rules from official sources. You have kept copies or written notes of these codes, and your planned setup operates entirely within them, including all mandated setbacks and limits on flock size.

Second: The yard map. You have mapped your yard for property lines, utility paths, drainage, and shade. You have selected a coop location that minimizes visual and odor impacts on your neighbors, and you have called your local 811 utility-locate service before planning to dig.

Third: The budget. You have calculated your startup costs, including the coop, run, feeders, and initial supplies. Just as importantly, you have estimated your ongoing monthly feed and bedding expenses, and set aside an emergency fund for unexpected issues, ensuring this project fits your household finances without strain.

Fourth: The flock choice. You have selected a sensible, legal number of laying hens. You have chosen climate-appropriate breeds based on traits rather than appearance, and you have made a deliberate choice between starting with chicks, pullets, or adult hens based on your risk tolerance and available setup time.

Fifth: The coop and run inspection. You have chosen or built a setup that is easy to access and clean. It is well-ventilated, weather-protected, and tightly secured with heavy-gauge hardware cloth, durable latches, and digging barriers to deter local predators.

Sixth: The household routines. You have scheduled exactly who will open the coop in the morning, who will close and lock it at dusk, who will check feed and water daily, and when the heavy weekly cleaning or maintenance will happen.

Seventh: The hygiene plan. You have established clear household rules for washing hands, keeping dedicated chicken-yard footwear out of the house, managing children and dogs around the flock, and handling, washing, and storing eggs safely to protect vulnerable people in your home.

Eighth: Emergency contacts. You have identified a poultry veterinarian, your local Extension office, or a state diagnostic lab. You have agreed on exactly what your household will do, and who you will call, if a bird becomes ill, injured, or dies suddenly.

Ninth: The neighbor plan. You have a strategy to manage manure, secure your feed against rodents, and control flies. You have a plan for how you will respectfully and constructively respond if a neighbor approaches you with a complaint.

Once you review your completed operating plan and this readiness check, you will arrive at one of three answers: yes, no, or not yet.

If your answer is not yet, normalize and celebrate that realization. Delaying your start is a massive success; it proves your planning process works. You might realize your current perimeter fence needs costly repairs, your household budget is a bit too tight this season, your work travel schedule is currently too unpredictable, or you simply need another month to properly level the ground for your coop. Choose the single largest barrier and solve it first, rather than trying to fix everything at once. Work on it at your own pace, and revisit the idea next season. A little nervousness is healthy, it proves you respect the responsibility of caring for live animals in a shared neighborhood.

If your answer is no, validate that decision. Suburban chicken keeping is a daily, year-round commitment that simply does not fit every lifestyle or property. If you have severe space limitations, strict HOA prohibitions, or realize the daily maintenance is not for you, walking away is a perfectly successful outcome. You can still enjoy the benefits of fresh food by supporting a local farmer, tending a backyard garden, or enjoying your neighborhood without the burden of daily livestock ownership. If your answer is yes, you are ready to move forward. Do so methodically. Buy your birds only after your coop and run are completely built, your predator-proofing is installed, your supplies are staged, and your transport is planned.

The goal of this journey is a small, legal, clean, contained, manageable, and neighbor-friendly flock. You have done the research. You have built the plan. The rest is simply making your decision, whenever you are ready.

Jennifer H. Johnson (Suburban Chicken Mom)

Appendix

Appendix

Your Worksheets

Appendix A. Local-law verification record

COMPLETE THIS VERIFICATION RECORD before you buy birds, order a coop, or purchase a single bag of feed. Print one copy for each source you need to contact.

It is common to hear that a neighbor keeps a flock without an issue, or to read a local social media post claiming rules recently changed. A cashier at the farm store might even say everyone buys chicks without getting a permit. However, informal advice from neighbors, online groups, or store employees is not reliable enough to base your decision on.

To avoid surprises and sunk costs, verify the rules directly with the proper authorities. Document the date, contact name, agency or role, answer, and source link for every conversation. Do not rely on your memory, and do not record phone conversations without permission. Instead, take detailed written notes on this form while you speak with officials or review documents.

Please note that completing this record supports your due diligence but does not constitute legal advice, guarantee legal compliance, or create official permission. If an answer is unclear, or if you live in a jurisdiction with overlapping authorities, request written confirmation or seek local legal counsel.

The U.S. local verification stack

You will likely need to complete a copy of this form for multiple overlapping authorities. A town might permit chickens, but your specific neighborhood association might ban them. Check each level of the stack that applies to your property:

- **Municipality:** City, town, or village code enforcement.
- **County/Township:** County health department, zoning board, or animal control.
- **Zoning District:** Your exact property zoning (e.g., R-1 Residential vs. A-1 Agricultural).
- **HOA/Deed:** Homeowners association bylaws, covenants, conditions, and restrictions (CC&Rs), or historical deed restrictions.
- **Lease:** Your landlord or property management company, if you are a renter.

Agency contact scripts

When you reach out to a local agency, remain polite, direct, and fact-focused. Do not argue about whether chickens make good pets; simply ask for the official ordinances.

Phone script: "Hello, my name is [Name] and I live at [Address]. I am researching whether it is permitted to keep a small flock of backyard hens on my property. Could you direct me to the exact ordinances or zoning rules that cover poultry, coops, and setbacks for my specific address?"

Email script: "To whom it may concern, I am writing to verify the local ordinances regarding backyard poultry for the property located at [Address]. Could you please provide the specific code sections or a link to the rules governing flock limits, coop setbacks, and any required permits for this zoning district? Thank you for your time."

Fictional completed example

To see how this looks in practice, review this fictional completed record.

- **Agency:** Springfield Municipal Government
- **Department:** City Zoning and Planning Department
- **Contact name:** David Chen, Permitting Specialist
- **Date:** March 14th
- **Exact answer received:** Hens are allowed, but a permit is required before building the coop. No roosters. Coops must be fully enclosed and kept clean to prevent odors and pests.
- **Source link/document:** Springfield Municipal Code Title 8, Chapter 4, Section B. Link: springfield.gov/code/poultry
- **Setback and structure rules:** Coops must be at least 15 feet from side and rear property lines, and at least 25 feet from any neighboring dwelling. Coops are classified as accessory structures and cannot be taller than 8 feet.
- **Flock limits:** Maximum of 6 laying hens on lots under half an acre.
- **Rooster rules:** Roosters are strictly prohibited within city limits.
- **Permit requirements:** Annual $25 backyard poultry permit required. Must submit a yard map showing coop setbacks with the application.
- **Waste and compost rules:** Manure must be kept in a sealed compost bin or removed weekly; cannot be stored within 20 feet of property lines or storm drains.
- **Egg sharing or sales rules:** Eggs may be consumed by the household or given away. Retail sales and roadside stands are prohibited in this zoning district.
- **Follow-up needed:** Print permit application, draft yard map showing the 15-foot property line setbacks, and mail the application with the $25 fee.
- **Next review date:** March 1st of next year (to renew the annual permit).
- **Status based on this source:** [X] Allowed [ ] Not allowed [ ] Not yet clear

Verification Record

Agency: (e.g., City Government, County Health, HOA Board)

Department: (e.g., Zoning, Animal Control)

Contact name: (Name and title of the person providing the information)

Date: Exact answer received: (Document the specific answer given regarding your address)

Source link/document: (URL, ordinance number, or physical document location)

Specific Rules Verified:

- **Setback and structure rules:** (Distances from property lines, houses, and max coop heights)
- **Flock limits:** (Maximum number of hens allowed, or square footage requirements per bird)
- **Rooster rules:** (Are males explicitly banned? Are there noise ordinances?)
- **Permit requirements:** (Initial application, inspections, fees, and renewal requirements)
- **Waste and compost rules:** (Rules for manure disposal, compost placement, or stormwater protection)
- **Egg sharing or sales rules:** (Are you legally allowed to sell eggs, or only keep them for personal use?)

Follow-up needed: (e.g., submit application, draw yard map, clarify a conflicting rule)

Next review date: (When permits expire or HOA rules are next updated)

Status based on this source: [ ] Allowed [ ] Not allowed [ ] Not yet clear

Keep these completed records stored safely with your approved permits, coop blueprints, and purchase receipts. Update the forms immediately if local rules change, if you move to a new property, or when an annual permit requires renewal.

My Notes

Appendix B. HOA, lease, and neighbor readiness checklist

Please note that completing this record supports your due diligence but does not constitute legal advice, nor does it guarantee approval. Allowed by your town does not automatically mean allowed by your homeowner's association, your lease, your property deed, or your household context. Attempting to keep your coop hidden to evade your lease or neighborhood covenants usually ends in community conflict and the risk of having to rehome your birds on short notice. Evasion is not a sustainable suburban strategy.

Use this worksheet to verify and document your specific private permissions before purchasing birds, coops, or supplies. Mark each section as Complete, Unresolved, or Not Applicable. If any required item remains Unresolved, pause your planning.

Part 1: Homeowner's Association (HOA) and deed restriction review

If you live in a deed-restricted community or a neighborhood with an HOA, your governing documents dictate what you can keep on your property. Do not rely on assumptions or neighborhood rumors.

[ ] **Document retrieval:** Locate and review your current Covenants, Conditions, and Restrictions (CC&Rs), declarations, deed restrictions, and published rules.

- Document name and date: ____
- Page numbers reviewed: ____

[ ] **Keyword search:** Check the rules specifically for the following terms: poultry, fowl, livestock, hens, roosters, pets, outbuildings, sheds, nuisance, noise, and animals.

- Notes on restrictions found: ____

[ ] **Address ambiguity:** If the covenants allow "customary household pets" but do not explicitly mention or exclude poultry, do not assume hens qualify. Get written clarification from the board.

- Date of inquiry: ____
- Contact name and role: ____
- Written answer received: ____
- Follow-up required: ____

[ ] **Final HOA approval:** Secure written confirmation from the community manager or board stating that a small flock of laying hens is allowed on your specific lot.

Part 2: Lease and landlord approval

If you rent your home, your lease terms and your landlord's permission control whether chickens are allowed. Standard residential leases rarely address backyard poultry explicitly; they usually limit traditional indoor pets.

[ ] **Lease review:** Read your current lease agreement, looking specifically for clauses regarding pets, outbuildings, property alterations, and yard use.

- Lease clause details: ______________________________

[ ] **Landlord conversation:** Pitch your plan clearly. Emphasize that you are requesting a small flock of laying hens (no roosters), housed in a contained run, with a strict waste-management plan.

- Date of conversation: ______________________________
- Landlord/Property manager name: ______________________________
- Initial verbal answer: ______________________________

[ ] **Written lease addendum:** Do not rely on a verbal "sure, that sounds fine." Secure a written lease addendum before spending any money. The addendum must specify:

- [ ] The exact maximum number of hens allowed.
- [ ] The approved location for the coop and run.
- [ ] Property cleanup and lawn repair expectations.
- [ ] Requirements for coop removal at the end of the lease.
- Date addendum signed: ______________________________

Part 3: Architectural and coop appearance approval

Many private covenants and leases treat a chicken coop as an outbuilding or exterior structure. This means it may be subject to formal architectural review to preserve neighborhood aesthetics.

[ ] **Guidelines reviewed:** Check community or landlord architectural guidelines for outbuilding, shed, and accessory-structure rules.

[ ] **Proposal prepared:** Gather the materials needed for formal approval:

- [ ] Exact coop and run dimensions.
- [ ] A site map proving the structure meets required property line setbacks.
- [ ] Paint colors and roofing materials (often required to match the primary residence).

[ ] **Formal approval:**

- Date submitted: ______________________________
- Contact name and role: ______________________________
- Approval received and filed: ______________________________

Part 4: Neighbor impact and boundary readiness

Neighbor readiness means having a physical and relational plan in place before the birds arrive. Do not wait until a neighbor complains about flies or wandering birds to figure out how you will manage your boundaries.

[ ] **Property line verification:** Verify actual property lines using a formal survey, property records, or plat map. Do not guess based on where a neighbor's lawnmower stops.

- Verification method: ____

[ ] **Physical containment plan:** Evaluate your perimeter fencing.

- [ ] Fence inspected for wide gaps, low heights, or weak gate latches.
- [ ] Confinement plan established (e.g., birds will be kept in a fully enclosed run rather than free-ranging to prevent wandering into neighbor yards).

[ ] **Odor and pest prevention plan:**

- [ ] Manure and soiled bedding compost/disposal site located securely away from neighbors' patios, windows, and outdoor living spaces.
- [ ] Feed storage utilizes secure, rodent-proof metal containers.

[ ] **Direct complaint-response channel:** Inform adjacent neighbors about your upcoming flock. Provide them with a direct way to contact you so that they bring concerns to you first, rather than immediately escalating to animal control or the HOA.

- Neighbor name (Left): ____
- Neighbor name (Right): ____
- Neighbor name (Rear): ____
- [ ] Contact method (phone number, text line, or email) provided to neighbors.
- [ ] Neighbors explicitly asked to reach out directly regarding noise, pests, or wandering birds.

Part 5: Household consensus

Readiness extends to the people living inside your own home. Ensure everyone is aligned on the daily realities of keeping a flock.

[ ] **Chore distribution:** All adults and participating household members agree on who is responsible for daily feeding, watering, and weekly coop cleaning.

- Primary caretaker: ____
- Backup caretaker: ____

[ ] **Financial agreement:** The household understands this is not a strategy for cheap groceries. A budget for secure housing, feed, bedding, and potential veterinary costs has been discussed.

[ ] **Contingency plans:**

- [ ] A vacation care and flock-sitting plan is outlined.
- [ ] Agreement reached on what happens if the chickens must be removed, rehomed, or if a chick turns out to be a rooster.

Walk-away triggers

Review your completed worksheet. Your situation is currently incompatible with backyard chickens if:

- Your HOA or deed explicitly bans poultry, livestock, or outside structures.
- Your landlord refuses to provide written permission or a lease addendum.
- Your lease prohibits outdoor structures or yard alterations.
- You cannot meet the required setbacks from your property lines.
- Your household is not in agreement regarding the daily chores, the budget, or the long-term commitment.

If you hit any of these walk-away triggers, stop. Do not attempt to hide a flock. Use this pause to reevaluate your housing situation, advocate for a rule change in your community, or wait until your circumstances change.

Appendix C. Agency call scripts and source log

Calling municipal offices, property managers, and homeowner associations can feel intimidating. However, city planners, code enforcement officers, and community managers vastly prefer answering questions before you build a chicken coop rather than responding to an angry neighbor's nuisance complaint later. Your goal is not merely to secure verbal permission. Municipal employees usually are not acting as your lawyer and may not be able to provide binding legal advice, but they can often direct you to the specific public ordinance, zoning code, or local bylaw that governs your property. Always ask for the source document so you can read the exact legal definitions for yourself.

The golden rule of documentation

As a default, do not audio-record conversations with agency staff, HOA board members, landlords, or neighbors unless you have confirmed the law in your jurisdiction and obtained any required consent. Recording-consent laws for telephone calls and private conversations vary by jurisdiction in the U.S., and unauthorized audio recording may violate wiretapping laws, creating serious legal risk. Furthermore, covert recording destroys community trust before your project even begins.

Instead, create a diligent manual paper trail. For every contact you make, use a dedicated notebook or digital spreadsheet to document the date, the contact name, the agency or role, the answer provided, the specific source link or code section cited, and any required follow-up.

Before you contact anyone

Gather the following information before you pick up the phone or send an email:

- **Your exact address and parcel number.** Zoning officers look up rules based on specific parcels, not just street names.
- **Your jurisdiction.** Verify whether you live within incorporated city limits or in an unincorporated county area. Unincorporated suburban properties are typically governed by county codes rather than municipal city codes.
- **HOA or management contact.** Have your homeowner association or property management contact information ready, if applicable.
- **Landlord contact.** Have your lease agreement and landlord contact information ready if you rent your home.
- **Your documentation system.** Have your source log open and ready to document the date, contact name, agency or role, answer, source link, and follow-up.

Script framework 1: Zoning and planning

Local zoning or planning departments regulate physical structures and land use. When you call this office, your focus should be on the coop structure, property lines, and required permits.

What to say: "Hello, I am verifying local codes for my residence at [Your Exact Address]. Could you tell me where I can find the current regulations for backyard chicken coops? I specifically need to verify the maximum flock size, the required setbacks from property lines for the coop and run, and whether this address requires a building permit or site-plan review for a small accessory structure."

The critical follow-up: If they give you a quick "yes" or "no," always ask: "Could you tell me the specific code section or direct me to the webpage where I can read that rule?"

Script framework 2: Animal control or code enforcement

While zoning handles the physical coop, animal control, environmental health, or code enforcement regulates the living birds, noise, and sanitation.

What to say: "Hello, I am checking local animal ordinances for my home at [Your Exact Address]. I am planning to keep a small suburban flock of laying hens. Can you direct me to the local rules regarding backyard poultry? I need to verify regulations regarding rooster prohibitions, noise or nuisance definitions, and daily waste disposal rules. If your office does not handle this, could you tell me which department does?"

The critical follow-up: "Thank you. Where can I find that ordinance online to ensure I stay in compliance?"

Script framework 3: Homeowner association or property management

HOA covenants and deed restrictions are often much stricter than municipal laws. Allowed by the city does not mean allowed by the HOA.

What to say: "Hello, I am a resident at [Your Exact Address]. I am reviewing our community guidelines regarding backyard setups. Before I make any plans, I need to know if our current covenants permit a small, contained chicken coop. If hens are permitted, could you point me to the exact section in the handbook that outlines the size limits, and tell me where to find the architectural request forms?"

The critical follow-up: "Thank you. Could you confirm which section of the bylaws I should review?" Always secure written approval from the architectural committee before purchasing materials.

Script framework 4: Landlords

If you rent, you must secure written permission from your landlord. Property owners primarily care about property damage, pests, and liability.

What to say: "Hello, I am interested in keeping a small, temporary, predator-resistant chicken coop in the backyard. I have verified that hens are legally permitted by the city at our address. I would only keep laying hens, no roosters, and I have a strict daily maintenance plan for keeping the area clean, odor-free, and rodent-proof. Would you be open to reviewing a short proposal outlining how I will manage the flock and protect the yard? If you approve, we can add a formal written addendum to the lease."

The critical follow-up: Follow up with a written email summarizing the maintenance plan and formally request a lease addendum.

Script framework 5: The local extension office

Many U.S. counties feature Cooperative Extension offices connected to land-grant universities. They do not enforce laws, but they provide localized agricultural guidance.

What to say: "Hello, I am a beginner planning to start a small backyard flock in [Your County]. Do you have any localized guides for backyard poultry keepers? I am looking for information on common local predators, Integrated Pest Management (IPM) guidelines for rodent control, any current biosecurity or avian influenza warnings in our area, and a list of local veterinarians who treat poultry."

The agency email template

If you prefer written answers, use this itemized template.

Subject: Ordinance inquiry for backyard hens at [Your Address]

Hello,

My name is [Your Name] and I reside at [Your Address]. I am planning to keep a small flock of backyard hens strictly for personal egg production, and I want to ensure my setup follows all local regulations.

Could you please provide the code sections, or a link to the ordinances, that address the following for my specific address and zone:

- The maximum number of hens permitted.
- Confirmation that roosters are prohibited.
- The required setbacks for an accessory agricultural structure from property lines, fences, and neighboring dwellings.
- Any building permits, site-plan approvals, or utility-locate (811) requirements for coop construction.
- Regulations regarding feed storage, pest control, and manure disposal.

Thank you for your time and guidance.

Best regards, [Your Name] [Your Phone Number]

The source log

Memory is a poor substitute for written records. Your personal source log allows you to point directly to municipal codes if your legally permitted flock is ever challenged. Use the columns shown in the completed example below. Log every single agency or office you contact. Document the date, contact name, agency or role, answer, source link, and follow-up. Save your fully completed source log alongside your coop plans, HOA approvals, and landlord addendums in your permanent household files.

Date	Contact name	Agency or role	Question asked	Answer given	Source link or code section	Follow-up
10/12	Jane Doe	City Planning Dept.	Accessory structure setbacks	Coop must be 10 ft from rear line, 5 ft from side.	Municipal Code § 15.4. Link emailed.	Submit site sketch for file. Done 10/15.
10/14	John Smith	Animal Control	Flock size and roosters	Max 6 hens. Roosters strictly prohibited.	Ordinance 8.04.020.	None required.

Appendix D. Yard map and coop placement worksheet

DRAW FIRST, DECIDE SECOND. Knowing your yard for weekend barbecues is entirely different from evaluating it as a year-round habitat and bio-management zone for a flock of laying hens. Use this layered mapping exercise and placement scorecard to compare at least two potential coop locations before committing to a final spot, buying a coop, or breaking ground.

The layered yard map

You do not need professional landscape design skills to complete this exercise. You can use the blank grid provided on the following page, print multiple copies of your property's satellite view from an online map, or use sheets of tracing paper over a hand-drawn perimeter.

Build your map in five functional layers to ensure no critical risk is overlooked.

Layer 1: Hard boundaries and existing footprint Begin with the physical realities of your lot.

- Draw the solid physical perimeter of your property. Add property-line notes indicating who owns which shared fence, if known.
- Outline your house, garage, existing sheds, patios, and decks.
- Mark existing gates and measure their width to ensure a wheelbarrow can pass through.

Layer 2: Legal limits and hidden utilities Apply the hard rules established during your local legal verification (Chapter 2).

- Draw dashed lines inward from your property lines to represent your municipal or county setbacks (e.g., 10 feet from the property line, 25 feet from a neighbor's dwelling). Cross out the "no-build" zones.
- **Utility marking reminder:** Before any U.S. digging related to coop, run, fence, post-hole, or predator-apron work, submit an 811 locate request by phone or through the state 811/one-call website and do not dig until the required waiting period and utility responses are complete. Confirm your state rules and consider a private locator for private lines that 811 may not mark. Once marked, draw these utility paths on your map. Avoid placing a coop directly over marked utility paths or recorded utility easements, because access may be required later; confirm placement restrictions with the utility, local locate service, or property records before building.

Layer 3: Water, weather, and topography Chickens cannot live in standing water, and mud breeds bacteria, flies, and odor.

- Mark any ground that stays squishy or muddy for days after a heavy rainstorm with an "X".
- Draw **runoff arrows** to indicate the natural direction water flows across your yard during a downpour. Never place a coop at the bottom of a runoff arrow.
- Track the sun's summer path. Mark intense, unprotected afternoon sun with an "S" and cool, deep shade with "SH".
- Draw an arrow indicating the direction of prevailing harsh winter winds.

Layer 4: Neighborhood and household zones Your coop must exist peacefully alongside your community and your family.

- Estimate and mark the location of neighbor windows, patios, and outdoor entertaining areas.
- Outline dog patrol routes, children's play sets, and high-traffic family areas.
- Mark vegetable gardens, delicate landscaping, or toxic plants to keep safe from scratching hens.

Layer 5: Access paths and candidate sites Identify at least two candidate locations (Site A and Site B) that survive the first four layers.

- Draw the physical footprint of the proposed coop and run.
- Draw an **access path** from your back door to the coop door. You will walk this path twice a day in freezing wind, heavy rain, and snow.
- Draw a chore route showing where a wheelbarrow will travel from the coop to your compost bin or yard-waste exit.

Map legend

Use the following symbols to mark up your grid.

- **Solid line** = Property line
- **Dashed line** = Legal setback boundary
- **Shaded zone** = Buried utility lines (811/local locate noted)
- **X** = Wet, muddy, or pooling area
- **Runoff arrow** = Direction of water flow
- **S** = Strong afternoon sun
- **SH** = Reliable afternoon shade
- **Dotted line** = Access path and wheelbarrow chore route
- **A / B** = Candidate coop sites

Final decision field

Review your scorecard. If Site A pools with water but has great daily access, it fails. If Site B is perfectly dry but sits three feet from your neighbor's patio chair, it fails the neighbor-impact test and likely the legal setback test. Comparing two sites forces you to think critically about the physical realities of your yard and prevents you from ignoring hidden flaws just because a spot looks picturesque.

Once you identify a site that passes the crucial tests, finalize your decision below. Keep this record handy when speaking to contractors, selecting a prefabricated coop, or buying building materials.

Selected site: ____

Backup site: ____

Deal-breaker problems found (if any): ______________________________

Utility marking confirmation number (811/local equivalent): ________________

Required groundwork (e.g., leveling, drainage correction, brush clearing):

____

____

Measurements to verify before buying or building: _______________________

Coop door should face (compass direction): ___________________________

Run should extend toward (compass direction): _________________________

Notes for contractor, installer, or designer:

____

____

Appendix E. Coop and run inspection checklist

Use this checklist before buying a prefabricated coop, building from plans, or moving birds into an existing structure. It is easy to fall in love with marketing photos. A freshly painted coop with scalloped trim and a shiny roof looks great on a website or sitting in the middle of a farm-store aisle. Because it is brand new, you might naturally assume it is functionally designed to house live animals and keep them safe.

Many commercially available backyard coops are built to ship easily in flat boxes and hit a specific retail price point, rather than to function well for chickens or the people caring for them. You must inspect function before trusting appearance. This tool helps you physically evaluate any setup to identify common functional flaws. While no structure is entirely predator-proof, used carefully, these criteria can make your setup more predator-resistant, safer, and easier to manage.

Evaluate your potential coop using three concrete action labels:

- **Pass:** The feature is technically sound and ready to use right now.
- **Fix First:** The feature is inadequate but can be securely modified (e.g., adding hardware, sealing a gap, or swapping chicken wire for hardware cloth) before birds arrive.
- **Fail (Walk Away):** The flaw is a fundamental deal-breaker that requires dismantling walls, rebuilding the core structure, or compromising your flock's safety.

Part 1: The Inspection Criteria

Review these technical requirements before completing the checklist.

Human Access: You must be able to reach all areas. If a frightened, injured, or sick bird is hiding in the farthest corner of the enclosed run or coop, you need to physically reach them without dismantling the roof. The main run access door must be wide enough to accommodate your shoulders, a feed bag, and a wheelbarrow.

Ventilation: Good airflow helps reduce the risk of respiratory illness, ammonia buildup, and frostbite. Openings must be positioned high above the birds' heads when they are sitting on the roosts, allowing warm, moist air to escape without blowing direct, cold drafts onto the flock.

Roosts: Chickens sleep flat-footed, not gripping like wild songbirds. Roosts should be built from wide lumber (like a two-by-four laid with the wide side up). They must be positioned higher than the floor of the nesting boxes; otherwise, hens will sleep in the nest boxes and soil the eggs.

Nest Boxes: You should be able to collect eggs without stepping foot inside the dirty run. Exterior nest box lids must be heavy enough to stay put in a strong gust of wind, or they require a latch. The back hinge must be sealed or covered so rainwater does not seep directly into the nesting material.

Cleaning Clearance: You must be able to maneuver a rake, hand trowel, or snow shovel inside the space without awkward twisting. If the coop uses a pull-out dropping tray, it must be deep enough (at least two inches) to hold bedding. If pulling the tray dumps dirty bedding into the wall cavities, the design is a failure.

Latches: Raccoons have excellent manual dexterity. Use latches that require multiple steps or opposing force to open, such as a carabiner, a padlock, or a two-step thumb release. Simple rotating wooden blocks or sliding bolts without a lock hole are inadequate.

Gaps: Inspect the doorframes, eaves, and floor joints. Predators like weasels and snakes can exploit surprisingly small openings. As a conservative predator-resistance target, seal or cover gaps larger than 1/2 inch.

Roof and Weather: The roof must overhang the walls far enough to direct heavy rainwater away from the side vents and the run flooring. The roofing material must be durable; thin, unpainted wood will warp and rot. The outdoor run should have at least a partially covered solid roof to provide a dry footprint during extended rain or snow.

Run Connection: The ramp leading from the coop down to the ground must attach securely without wobbling. The small chicken pop door must be lockable from the outside so you do not have to crawl into the run to secure it at night. The seam where the wire run meets the wooden coop siding must be flush.

Hardware Cloth: Chicken wire may keep chickens in, but it is poor protection against many predators. For a predator-resistant build, cover vents, windows, and run panels with 1/2-inch welded hardware cloth. Secure it with robust fasteners, such as screws and washers or heavy-duty poultry staples; do not rely on lightweight staples. If you are installing a hardware cloth apron or skirt around the perimeter to deter digging predators, submit an 811/local utility-locate request before breaking ground; follow your state or local waiting period and remember that private lines may require a private locator.

Maintenance Clearance: Heavy wire mesh panels should be mounted to the outside of the framing so they can be easily replaced if damaged, rather than sandwiched irretrievably between glued panels. You must be able to easily unscrew or lift out the roosting bars to scrub them outside during an annual deep clean or a flock mite treatment.

Part 2: The Field Checklist

Take this checklist with you when inspecting a fully assembled display model, evaluating an existing structure, or reviewing a manufacturer's design specifications.

Human access [ ] Pass [ ] Fix First [ ] Fail: I can reach all corners, catch a hiding bird, and enter with supplies.

Ventilation [ ] Pass [ ] Fix First [ ] Fail: Vents are positioned high above the roosts to release moisture without causing drafts.

Roosts [ ] Pass [ ] Fix First [ ] Fail: Roosts are wide, flat, and positioned higher than the nest boxes.

Nest boxes [ ] Pass [ ] Fix First [ ] Fail: Accessible from the outside with weather-proof, latched hinges.

Cleaning clearance [ ] Pass [ ] Fix First [ ] Fail: Shovel/rake fits easily, and dropping trays (if used) do not spill into walls.

Latches [ ] Pass [ ] Fix First [ ] Fail: Exterior doors use two-step latches, carabiners, or padlocks.

Gaps [ ] Pass [ ] Fix First [ ] Fail: Eaves, frames, and seams meet the conservative 1/2-inch-or-smaller gap target.

Roof and weather [ ] Pass [ ] Fix First [ ] Fail: Roof overhangs vents, material is waterproof, and run has a dry zone.

Run connection [ ] Pass [ ] Fix First [ ] Fail: Ramp is sturdy, pop door locks from the outside, and run seams are flush.

Hardware cloth [ ] Pass [ ] Fix First [ ] Fail: Openings use 1/2-inch welded hardware cloth, not chicken wire, secured with robust fasteners.

Maintenance clearance [ ] Pass [ ] Fix First [ ] Fail: Roosts can be unscrewed for cleaning, and mesh panels are accessible.

Part 3: Completed Sample Audit

To see how this decision matrix works in practice, consider this sample audit of a popular, $400 flat-pack kit coop sold online.

- **Appearance:** Painted red wood with crisp white trim and a small attached run.
- **Roosts:** The primary roosting bar sits an inch lower than the entrance to the nesting boxes.
- Score: Fix First. You must purchase a piece of two-by-four lumber and install a higher bar before birds arrive, or the hens will sleep in the boxes and soil the eggs.
- **Hardware cloth:** The coop features large, high vents, but they are covered with thin chicken wire held in by lightweight office staples.
- Score: Fix First. You would need to purchase heavy-duty 1/2-inch hardware cloth, cut it to size, and secure it over the vents using screws and fender washers or another robust fastening method.
- **Cleaning clearance:** The slide-out dropping tray is less than one inch deep. The human access door is only twelve inches wide.
- Score: Fail (Walk Away). A standard shovel cannot fit inside, and the impossibly shallow tray means dirty bedding will scrape off and dump directly into the internal wall cavities every time you pull it out. You cannot safely fix this without tearing the coop apart. Walk away from the purchase.

Do not move birds into a coop until every category registers as a solid Pass. Replace unacceptable hardware, cover all gaps, and adjust roost heights before the flock arrives.

Appendix F. Predator-resistance audit

THIS AUDIT REDUCES COMMON risks; it does not guarantee complete safety. No outdoor structure is entirely predator-proof. Many beginners assume their neighborhood is completely clear of threats. But raccoons use suburban storm drains, hawks sit on streetlamps, foxes navigate greenbelts, and household dogs have strong prey drives. The predators are already present; they just have not visited your yard yet.

This tool helps you physically evaluate any prefabricated coop, custom build, or existing shed to identify common functional flaws. By applying these specific inspection criteria, you remove easy opportunities for predators to enter. Local threats vary, so consult your county Extension office, state wildlife agency, or local vector-control board about specific risks in your area.

How to use this audit: Wait until dusk. Bring a bright flashlight, a phone, and a pencil. Turn the flashlight on inside the dark, closed coop. Any beams shining outward will immediately highlight hidden structural gaps. Walk the perimeter from the ground up, inspecting closely. Photograph weak points before leaving the area so you know exactly what hardware to buy.

1. Digging predators (Dogs, foxes, coyotes, and rats)

Predators will aggressively dig under the edges of a run to gain access. You must inspect the ground-level perimeter where the enclosure meets the earth.

- **The apron/skirt:** Verify the presence of a physical digging barrier. This is typically a hardware-cloth apron extending outward 18 to 24 inches from the base of the run.
- **Installation check:** Ensure the skirt is either securely pinned flat against the grass so vegetation grows up through it, or properly trenched into the soil.
- **Digging warning:** If you plan to dig or trench a new apron into the ground, submit an 811 locate request in the U.S. by phone or through the state 811/one-call website, or use your local utility-locate equivalent. Do not dig until the required waiting period and utility responses are complete, and consider a private locator for private lines that 811 may not mark.
- **Erosion:** Check for soil that has washed away under the frame, or uneven ground creating gaps beneath the baseboards. Look for signs of wild animal scratching or eroded dirt.

Status: [ ] Pass [ ] Needs fix [ ] Photo taken

Priority: [ ] Today [ ] This week [ ] Monitor

Repair needed: ____

Date fixed: ______________

2. Reaching and squeezing predators (Raccoons, weasels, snakes, and mice)

Raccoons are notorious for reaching through wire to grab resting birds, while weasels and mice can squeeze through impossibly small holes.

- **The mesh check:** For predator resistance, verify that run walls and ventilation windows are covered in half-inch welded hardware cloth. Lightweight, hexagonal chicken wire may keep chickens contained, but it is not a reliable predator barrier. Dogs and raccoons can tear, bend, or bite through basic chicken wire.
- **The half-inch rule:** Inspect all structural openings. Treat any gap larger than half an inch as a priority weak point and cover or seal it.
- **The flashlight test:** Check the exterior while the flashlight shines brightly inside the enclosed coop. Note every spot where light escapes around vents, windows, nesting box hinges, and the roofline.
- **Mesh condition:** Press firmly against mesh panels to ensure there is no bowing. Check for worn galvanized coating or rust spots that compromise the wire's strength.

Status: [ ] Pass [ ] Needs fix [ ] Photo taken

Priority: [ ] Today [ ] This week [ ] Monitor

Repair needed: ____

Date fixed: ________________

3. Climbing and prying predators (Raccoons and opossums)

Raccoons have excellent climbing abilities, extreme persistence, and dexterous front paws that function like hands. They will pry at loose edges and manipulate basic locks.

- **Latches:** Apply the toddler test. If an inquisitive two-year-old could open the latch, a raccoon will eventually figure it out. Photograph any latch that can be opened in a single motion, such as a simple hook-and-eye or a basic sliding barrel bolt.
- **Nighttime lockup:** Verify that every single door holding chickens at night is secured with a two-step locking mechanism. Acceptable options include a carabiner clipped through a sliding latch, a padlock, or a heavy spring-loaded thumb-release latch.
- **Fasteners:** Inspect how the mesh is attached to the wood. Do not rely on light-duty staple guns. Verify the use of secure hardware, such as heavy-duty pneumatic poultry staples or screws paired with wide washers.
- **Corners and seams:** Walk to every high-stress corner where wire panels meet the frame. Look for wood that has warped, split, or rotted, leaving the sharp edge of the wire exposed and vulnerable to prying.

Status: [ ] Pass [ ] Needs fix [ ] Photo taken

Priority: [ ] Today [ ] This week [ ] Monitor

Repair needed: ____

Date fixed: ________________

4. Flying and dropping predators (Hawks, owls, and climbing raccoons)

Threats come from above, both during the day and at night. Open-topped runs leave flocks highly vulnerable.

- **Roof and top cover:** Check the overhead enclosure. Note that while lightweight bird netting successfully deters daytime hawks, it will rapidly fail under the weight of climbing raccoons, chewing squirrels, heavy snow accumulation, or falling tree branches. Hardware cloth or a solid roof is vastly superior.

- **Eaves and rafters:** Inspect the eaves on solid-roofed coops. Verify that the open spaces between the roof rafters are tightly sealed with hardware cloth to stop climbing predators from squeezing over the top of the walls.

- **Damage check:** Check all existing overhead wire or netting for snags, drooping sections, and small holes.

Status: [ ] Pass [ ] Needs fix [ ] Photo taken

Priority: [ ] Today [ ] This week [ ] Monitor

Repair needed: ____

Date fixed: ________________

5. Openings and the nighttime connection

The hardware is only effective if the architectural joints remain tight and the doors sit flush.

- **The coop-to-run joint:** Inspect the transition point where the enclosed wooden coop meets the wire run. Coops and runs settle into the dirt at different rates. Over time, this settling frequently opens hidden, wedge-shaped gaps between the two structures.

- **The pop door:** Check the small door that allows chickens to move between the coop and the run. Verify that it drops fully into a channel or locks securely into place at night so it cannot be lifted easily from the outside with a paw.

- **Human access doors:** Close the human access door and the coop clean-out doors. Inspect door frames for seasonal warping caused by humidity or freezing temperatures. Every door must sit flush against the frame without noticeable gaps at the top or bottom.

Status: [ ] Pass [ ] Needs fix [ ] Photo taken

Priority: [ ] Today [ ] This week [ ] Monitor

Repair needed: ____

Date fixed: ________________

6. Feed attractants and the dog buffer

Predator management relies heavily on limiting what attracts wild animals to your yard in the first place, and keeping familiar animals at a safe distance.

- **Feed spills:** Look at the ground directly beneath your feeders. Massive amounts of spilled grain attract mice and rats. A booming rodent population will subsequently attract snakes, foxes, and larger predators to your yard.
- **Storage containers:** Inspect your feed storage. Bags of feed left on the ground in a garage or plastic deck box invite chewing rodents. Store feed in heavy, chew-resistant metal bins with tight-fitting lids.
- **The household dog barrier:** Check the line of sight between your flock and any nearby dogs, including family pets. Look for areas where a jumping dog could hit the wire mesh, causing panic to the flock and structural damage to the run. Ensure there is a buffer zone to prevent your dog from pacing the immediate perimeter.

Status: [ ] Pass [ ] Needs fix [ ] Photo taken

Priority: [ ] Today [ ] This week [ ] Monitor

Repair needed: ____

Date fixed: ______________

After the audit Before you walk away, choose the most urgent structural flaws from your list and schedule them for immediate repair. Never leave a known nighttime access gap for later. Take your list to the hardware store, make the repairs, and repeat this audit after any major storm, period of high wind, or attempted predator entry. Ensure your enclosure remains a durable, resistant barrier for the lifespan of your flock.

Appendix G. Starter flock decision tables

Use these tables after your local legal rules, homeowner's association guidelines, and available yard space are fully verified. A backyard flock is a living biosecurity system. Mixing too many different sizes, temperaments, life stages, and farm sources in a small suburban setup creates stress, bullying, and significant disease risk. Choose fit over novelty to build a harmonious flock that suits your climate, budget, and actual daily schedule.

Table 1: The life stage decision

Chicks (0 to 6 weeks old)

- **Best fit:** Keepers with a safe, draft-free, temperature-controlled indoor space who want to manage the brooding phase.
- **Equipment and labor:** High. Requires an indoor brooder, a fire-safe heat plate, chick-specific feeders, and daily hygiene and temperature monitoring.
- **Egg timeline:** Long. You will feed them for five to six months before seeing a single egg.
- **Rooster risk:** High. Sexing errors happen, even with highly reputable commercial hatcheries.
- **Source criteria:** Buy from hatcheries participating in the National Poultry Improvement Plan (NPIP). Understand that NPIP certification means the flock is tested for specific illnesses like Pullorum-Typhoid and Avian Influenza; it does not mean the birds are immune to all diseases. Request Marek's disease vaccination prior to shipping, as it is difficult to administer at home.

Point-of-lay pullets (16 to 24 weeks old)

- **Best fit:** Most suburban beginners. Pullets skip the messy, dust-heavy indoor brooding phase.
- **Equipment and labor:** Medium. Fully feathered pullets bypass the brooder and go directly into your secure outdoor coop and run.
- **Egg timeline:** Short. Depending on moving stress, breed, and daylight, they usually begin laying within two to four weeks.
- **Rooster risk:** Lower than chicks, but accidental roosters still occasionally occur.
- **Source criteria:** Ask for vaccination records. Buy your entire starter flock from one single source. Sourcing pullets from three different farms and immediately mixing them in your backyard is the fastest way to trigger a fatal respiratory illness.

Adult hens (1 year or older)

- **Best fit:** Keepers adopting from rescues or farm downsizings who prioritize simplicity over maximum egg production.
- **Equipment and labor:** Low. They move straight into the outdoor setup, though older birds may take longer to adapt to new routines.
- **Egg timeline:** Immediate to short. Moving stress or a seasonal molt can temporarily stop laying.
- **Rooster risk:** Zero. You can visually confirm the bird is a hen.
- **Source criteria:** Adult birds can introduce latent diseases, especially when added from outside flocks. If you are starting fresh, adopt the whole group together. If you ever add an adult bird to an already existing flock, keep any new or returning birds physically separated from your resident flock for at least 30 days before mixing them. Use a separate enclosure, separate feeders and waterers, and dedicated or disinfected tools, clothing, and footwear; care for the resident flock first, then the quarantined birds. If respiratory disease risk is a concern, ask a poultry veterinarian or Extension specialist how far apart the enclosures should be and how to manage airflow. Quarantine reduces risk, but it does not guarantee that every infectious disease, parasite, or carrier state will be detected.

Table 2: Flock size templates

Manufacturer capacity claims on prefabricated coops are often dangerously exaggerated. Verify the square footage yourself. For suburban laying hens, plan for an absolute minimum of 4 square feet of interior coop space and 10 square feet of run space per bird, though more is always better.

Never max out your legal limit by default; you must leave legal and physical room to separate a bullied or injured bird later. The egg numbers below are peak-season, first-year planning estimates. They are not guarantees. Production drops significantly during extreme heat, extreme cold, shorter winter days, and seasonal molts.

The 3-hen starter flock

- **Egg planning estimate:** 12 to 15 eggs weekly.
- **Space pressure:** Lowest.
- **Flock dynamics:** A practical minimum number for maintaining a healthy social structure. If one bird is lost to illness or a predator, the remaining two still have each other and avoid isolation stress.

The 4-hen standard flock

- **Egg planning estimate:** 16 to 20 eggs weekly.
- **Space pressure:** Moderate.
- **Flock dynamics:** Highly stable. This size allows for a natural pecking order without overcrowding the enclosure, making it the best fit for most suburban beginners.

The 6-hen maximum starter flock

- **Egg planning estimate:** 24 to 30 eggs weekly.
- **Space pressure:** Significant.
- **Flock dynamics:** Active and busy. Requires a noticeably larger physical footprint, aggressive waste management, and plenty of visual barriers in the run so lower-ranking birds can escape dominant top hens.

Table 3: Breed trait priorities

Look for specific traits that match your physical environment. No hatchery can guarantee an exact personality or egg count, but genetics provide a strong baseline.

- **Climate tolerance:** Prioritize your hardest season. Cold-hardy breeds typically have heavier bodies and smaller combs that resist frostbite. Heat-tolerant breeds are usually lighter with larger combs to dispel body heat. Manage your secondary season with shade, ventilation, dry bedding, or wind protection.
- **Temperament:** Heavy breeds are often more docile, calmer around children (with mandatory adult supervision), and tolerate confinement well. Lighter breeds are highly active foragers and very predator-aware, but they can be flighty, skittish, and noisy.
- **Egg expectations:** Production hybrids are bred for maximum daily output but carry a much higher risk of reproductive illness later in life. Heritage breeds lay fewer eggs per week but generally offer better longevity and steady hardiness.

The rooster contingency plan

Suburban beginners should strongly avoid roosters due to noise ordinances, neighbor conflict, and potential aggression. Because sexing errors occur even with top hatcheries, you must have a plan for an accidental male before you buy chicks or pullets. Do not release a rooster into a park, woods, or rural road; abandonment is often illegal, unethical, and a death sentence for the bird. Check one confirmed plan:

- [ ] I am buying pullets from a local seller who provides a written return or trade-in policy for accidental roosters.
- [ ] I have confirmed directly with my local animal control shelter that they legally accept roosters.
- [ ] I have spoken directly to a local farm or sanctuary, and they have explicitly agreed to accept unwanted roosters.

Completed example: The Miller family

To see how these tables work together, imagine a family with two young children, a medium-sized yard, and cold Midwestern winters.

- **Life stage:** They choose **pullets (16 to 24 weeks)** to bypass the fire hazards and dust of an indoor brooder.
- **Flock size:** They choose **4 hens**, ensuring their planned custom coop easily exceeds the 16-square-foot interior minimum.
- **Breed priorities:** They prioritize **cold-hardy** and **docile** traits. They select Buff Orpingtons, noting they must still actively supervise their children around the flock.
- **Source criteria:** They buy all four pullets from one single farm to avoid mixed-source disease risks, verifying the farm vaccinates for Marek's disease.
- **Rooster plan:** They secure a written return guarantee from the seller.

My short-list worksheet

Use this worksheet to lock in your short list. If your selected choices do not fit your legal rules, verified space, climate, and rooster plan, revise them before you buy.

- **My life stage:** ____________________________
- **My flock size:** ____________________________
- **My climate priority:** ____________________________
- **My temperament priority:** ____________________________
- **My single-source supplier candidate:** ____________________________
- **My rooster plan:** ____________________________
- **My 2–3 breed candidates to research:** ____________________________

Appendix H. First-year budget worksheet

LOOK DOWN INTO A farm store trough full of peeping day-old chicks, and it is easy to think: Chicks are cheap. At just five or six dollars a bird, the initial purchase feels like spare change. But the bird itself is the least expensive part of backyard chicken keeping. You are not just buying a pet; you are building, supplying, and maintaining an outdoor residential biosecurity and containment system.

To avoid financial surprises, remember this key takeaway: budget for the flock system, not just the birds. Do not expect backyard hens to save money on groceries.

Modeled first-year assumptions

Because local prices vary wildly based on your region, inflation, and local material costs, a blank budget tool is often too abstract to be useful. To help you anchor your expectations, review these three modeled scenarios for a standard four-hen suburban starter flock.

Scenario 1: Minimum legal and safe ($800–$1,200) This is a high-labor, lower-capital approach. It assumes you are repurposing an existing shed or building a small coop from scratch using salvaged materials that have been carefully vetted for hazards such as lead paint and rot. You will still spend heavily on new, heavy-duty hardware cloth and secure latches to improve predator resistance. This budget assumes you are purchasing basic galvanized feeders, standard commercial layer feed, and setting aside a minimal emergency fund. It requires you to do the manual labor of trenching the run. Before digging, trenching, or installing buried fencing, contact 811 or submit a locate request through your state 811 center, then wait the legally required time for utility responses and markings. Check state/local rules and arrange private locating for owner-installed lines that 811 does not cover. It also requires building your own secure access points.

Scenario 2: Comfortable beginner ($1,500–$2,500) This is the most common suburban entry point. It assumes you are purchasing a mid-range prefabricated coop or kit, but you are also budgeting the necessary funds to retrofit it. Most prefabs require an additional $150 to $200 in predator-resistance upgrades, including hardware-cloth skirts, heavy-duty carriage bolts, and carabiner latches. This tier includes purchasing a safe indoor radiant brooder plate for chicks, dedicated metal feed storage bins, quality commercial feed, and establishing a baseline veterinary diagnostic fund.

Scenario 3: Premium upgrade ($3,000+) This tier trades capital for convenience and longevity. It assumes a custom-built, walk-in coop designed for human maintenance clearance, a fully covered and trenched run, and an automated, predator-proof coop door. This budget incorporates premium organic or locally milled pelleted feed, automated heavy-duty waterers, high-end construction materials that match the primary house, and a robust emergency veterinary reserve.

Completed sample: The comfortable beginner

Below is a modeled breakdown of what the "Comfortable beginner" tier actually looks like on paper for a four-hen flock during its first year. **Note: These are sample reference numbers to show how the budget functions. You must verify actual prices in your zip code.*

Startup infrastructure (One-time costs)

- **Coop and run kit:** $750
- **Predator-resistance upgrades:** $160 (Two 10-foot rolls of 1/2-inch hardware cloth, washers, screws, and four padlocks)
- **Adult feeders and waterers:** $65 (Durable, weather-resistant plastic or galvanized metal)
- **Secure storage:** $70 (Two 30-gallon galvanized metal trash cans with tight-fitting lids)
- **Dedicated cleaning tools:** $45 (Sturdy metal rake, scraping shovel, stiff brush, utility gloves)
- **Brooder supplies (Conditional):** $130 (Large indoor tub, radiant heat plate, chick-sized feeder/waterer)
- **Permits (Conditional):** $50 (Municipal registration fee)
- **The birds:** $25 (Four sexed pullet chicks)

Recurring and contingency costs (First year)

- **Feed:** $216 (Calculated at 360 total pounds for four birds over one year, at $0.60 per pound)
- **Bedding:** $120 (Pine shavings for the coop, refreshed regularly)
- **Veterinary and diagnostic fund:** $200 (Baseline for one state lab swab or basic exotic vet exam)
- **Replacement and repair fund:** $100 (For rusted latches, winterizing material, or patching wire)
- **Emergency fund:** $150 (Contingency for unexpected infrastructure failure, like weather damage)
- **Total modeled first-year cost:** $2,081

Your local price-check directory

Do not base your financial readiness on national averages or the sample above. Before you purchase birds, call your local farm stores, hardware suppliers, and county offices to fill out the blank worksheet in your companion PDF. You must research the following local line items:

1. Feed price per pound Do not budget by the bag; budget by the pound. A flock of four laying hens will consume roughly 360 pounds of feed in their first year (averaging 1/4 pound per bird, per day, plus unavoidable waste). Formula: Bag price ÷ Bag weight = Price per pound. Example: A $24 bag weighing 40 pounds is $0.60 per pound. Multiply your local price per pound by 360 to find your annual base feed cost for a four-hen flock.

2. Hardware cloth and lumber Check the current price of a 50-foot roll of 1/2-inch, 19-gauge hardware cloth. Do not treat chicken wire as predator-resistant for a suburban coop or run; price out 1/2-inch welded wire hardware cloth or another stronger welded barrier appropriate to your local predator pressure. Check the current cost of exterior-grade 2x4s and paving stones if you plan to build a predator skirt.

3. Veterinary and diagnostic fund Call your state university Extension office, state veterinary diagnostic lab, or a local exotic animal veterinarian. Ask for the baseline cost of a basic poultry exam or a diagnostic swab for a suspected illness. Enter this exact number into your budget as a non-negotiable cash reserve.

4. Replacement, repair, and emergency funds You must set aside a specific cash line item for infrastructure failure. Roofs leak, latches rust, waterers crack in freezing temperatures, and predators test fences. Budgeting at least $100 to $200 for repairs prevents a minor structural failure from becoming a fatal security breach simply because you lacked the funds to fix it immediately.

5. Vacation and care coverage If you travel, you must budget for care. Call local pet sitters or talk to neighbors to determine the daily rate for someone to check the birds, refresh water, collect eggs, and verify that the coop is securely locked at dusk.

Final affordability check

Once you have filled out your local worksheet in the companion PDF, add your subtotals to find your total estimated financial commitment for the first year.

Ask yourself one question: Can I pay this total without cutting safety, biosecurity, or predator-resistance corners?

Do not use hoped-for egg savings to justify an unsafe or unaffordable setup. If your answer to the final question is no, this is a responsible "not ready yet" outcome. Save up, revise your infrastructure plan, or delay the flock by a full year until you can build a secure, predator-resistant system from day one. Keep your final budget total strictly in mind as you gather the supplies that will safely bring your birds home.

Appendix I. Supplies and shopping list

Walking into a farm supply store with an empty cart and a new flock on the way can quickly lead to impulse buying. Before you purchase decorative extras, focus on your essential biosecurity, containment, and daily care tools. You might assume a pre-packaged starter box covers all your bases, but these kits rarely include everything required for a secure setup and often feature small feeders that only work for the first few weeks of a bird's life.

Use the segments below to build a targeted shopping list. Focus your initial budget on the required biosecurity and safety infrastructure, and save the optional upgrades for later.

Required before arrival: Universal supplies

Whether you are starting with day-old chicks, feathered pullets, or adult hens, these items must be purchased, assembled, and placed before your birds arrive on the property.

- **Secure feed storage:** Buy a heavy-duty bin with a tight-fitting, latching lid. Metal containers are highly recommended if chewing rodents are a known issue in your neighborhood.
- **Dedicated cleaning tools:** Purchase a heavy-duty plastic bucket, a stiff-bristled scrub brush, and a plastic scraper. Label these and use them exclusively for coop chores to keep poultry manure away from household cleaning supplies.
- **Household hygiene station:** Set up your backyard boundary. Purchase dedicated waterproof coop boots that stay outside. Station outdoor hand soap or hand sanitizer near your exterior hose or coop entrance to ensure immediate handwashing.
- **Secure transport carrier:** Obtain a hard-sided plastic pet carrier with a secure locking door. Do not rely on cardboard feed store boxes for anything other than the initial ride home for day-old chicks. A proper carrier is necessary for isolating an injured bird, safely containing a bird for observation, or evacuating during an emergency.
- **Flock records folder:** Set up a physical folder or digital file to hold your legal verification record, HOA approval, utility-locate confirmation, feed store receipts, seller vaccination records, and local emergency contact numbers.
- **Bulk bedding:** Secure your run and coop bedding based on your maintenance plan. Coarse construction sand, large-flake pine shavings, or hemp bedding are common choices depending on local availability and your composting setup.

Safety-critical infrastructure

Do not compromise on items related to fire prevention, electrical safety, or predator resistance. Substandard materials here can lead to flock loss or property damage.

- **Hardware cloth:** If you are upgrading a prefabricated coop or building a custom run, purchase 1/2-inch welded wire hardware cloth. Do not buy flexible, hexagonal chicken wire, which raccoons and dogs can easily tear open.

Include heavy-duty washers and screws for secure mounting.

- **Outdoor-rated electrical supplies:** If you are running temporary power to a heated waterer during winter, verify that all extension cords are heavy-duty, specifically outdoor-rated, and completely protected from weather, foot traffic, and chewing rodents.
- **GFCI protection:** For outdoor outlets powering poultry equipment, use GFCI-protected outlets and weatherproof covers consistent with applicable electrical code and manufacturer instructions. Check your existing outlets before plugging in equipment. Do not attempt DIY electrical wiring; consult a licensed electrician if your setup requires a new permanent power source.

Chick-only supplies

If you are brooding day-old chicks indoors or in a garage, you need specific temporary infrastructure.

- **Draft-free brooder enclosure:** Use a galvanized stock tank, commercial brooder, or large, heavy-duty plastic storage tote with high, solid sides to block drafts. It must have a secure wire-mesh top to enforce a strict boundary against household pets, cats, and small children. Do not place the brooder in a kitchen, dining room, or food-prep area.
- **Safety-verified heat source:** Radiant heat plates are highly recommended for suburban brooders because they mimic a mother hen and can pose a lower fire risk than traditional heat lamps when used as directed. If you must use a heat lamp, secure it with a chain or other secondary restraint, avoid relying solely on the attached spring clamp, and keep it well away from bedding and other flammable materials. Follow all manufacturer instructions carefully and test the setup for a full day before chicks arrive.
- **Brooder bedding materials:** Stock plain paper towels to line the brooder floor for the first few days. This provides necessary traction for developing legs and helps chicks locate scattered feed. Transition to standard pine shavings later. Avoid cedar shavings, which may irritate fragile respiratory systems.
- **Chick starter feed:** Purchase feed specifically labeled "chick starter," which comes in a fine crumble that tiny beaks can easily digest.
- **Chick-sized feeder and waterer:** Buy shallow chick waterers to minimize drowning risks, and narrow feeders to keep chicks from walking and sleeping in their food.

Pullet and adult-only supplies

Older birds moving directly into your outdoor coop require permanent backyard equipment.

- **Stage-appropriate feed:** Rapidly growing pullets require a transitional "grower feed." Actively producing hens require calcium-rich "layer feed." Check the labels carefully in the feed aisle to ensure you are matching the diet to your flock's exact life stage.
- **Adult-sized feeder and waterer:** Invest in heavy-duty galvanized metal or UV-resistant plastic models. Select designs that can hang securely from the run ceiling or sit elevated on concrete blocks to keep bedding, dirt, and manure out of the trays.
- **Nesting box materials:** Stock soft materials to encourage laying in the correct spot and to prevent cracked eggs. Clean straw, thick pine shavings, or commercial pre-cut nesting pads work well.
- **Egg collection basket:** Keep a dedicated container for gathering eggs each morning, such as a small wire basket, a padded bucket, or a designated plastic bin that can be easily sanitized.

- **Unwashed egg storage:** Establish your household egg-handling supplies. If your household default is to leave unwashed, clean-shelled eggs at room temperature, purchase a dedicated countertop skelter or specific reusable cartons that are kept strictly separate from store-bought, washed eggs.

Do-not-buy-yet and optional later items

Delay the following purchases to protect your initial budget and prevent immediate complications.

- **Automated coop doors:** Wait until you have established a daily routine, observed your flock's natural sleep schedule, and thoroughly inspected your run for predator gaps before relying on automation to lock your birds up at night.

- **Treats and scratch grains:** New arrivals of any age should focus entirely on their nutritionally complete starter, grower, or layer feed. Wait at least a month before introducing treats or scratch grains, and limit them strictly to prevent digestive upset and behavioral issues like feather pecking.

- **Toys and decorative signs:** Chicken swings, xylophones, and wooden coop signs are not required before arrival. Put this money toward biosecurity hygiene stations, high-quality hardware cloth, or establishing your emergency veterinary fund.

- **Over-the-counter medications:** Do not preemptively buy farm-store antibiotics, dewormers, or complex water additives. You should stock basic first-aid supplies like saline eye wash and poultry-safe wound spray, but avoid diagnosing, prescribing, or dosing sick birds on your own; seek direct guidance from a poultry veterinarian, Extension specialist, or your state diagnostic lab.

Appendix J. Chick, pullet, and adult hen arrival checklists

Complete the relevant checklist before you leave the house to pick up your birds or before the post office calls for delivery. Beginners often assume they can finish assembling the brooder or spreading the final layers of coop bedding while their new flock waits in a cardboard transport box. Resist this temptation.

Transport exposes birds to motion, noise, temperature shifts, and unfamiliar handling. When they arrive at your property, they need to transition immediately from their travel container into a quiet, fully prepared environment. Attempting to hammer a loose piece of hardware cloth, adjust a heat source, or fill a waterer while the birds wait adds unnecessary confusion to a sensitive day. Your goal is a calm, predictable transition.

Choose the pathway below that matches the age of the birds you are bringing home. Before leaving, complete the pre-departure baseline steps regardless of which age group you are acquiring.

The pre-departure baseline

- [ ] **Emergency contacts saved.** Store your avian or farm-friendly veterinarian's phone number, the local agricultural Extension office contact, and the seller's information in your phone.
- [ ] **Store hours checked.** Note the closing time of your nearest farm supply store in case a waterer cracks or a brooder heat lamp bulb burns out on the first evening.
- [ ] **Do-not-transport rule acknowledged.** If you suspect you are picking up sick birds, or if your newly arrived birds show immediate signs of contagious disease, sudden death, or multiple illnesses, do not load them back into your car to seek help. Call your veterinarian, diagnostic lab, or Extension office for instructions first.

Pathway 1: The day-old chick arrival checklist

Newly arrived chicks cannot regulate their own body temperature. They require precise, stable warmth, immediate access to room-temperature water, and strict household hygiene boundaries.

- [ ] **Test the brooder temperature.** Turn on your heat plate or heat lamp at least twenty-four hours before arrival to verify the temperature is stable. Confirm the heat source is safely secured away from pine shavings and flammable materials.
- [ ] **Block all drafts.** Verify the brooder is positioned away from exterior doors, drafty windows, and direct streams from air conditioning vents.
- [ ] **Establish household hygiene boundaries.** Ensure the brooder is not in a kitchen, dining room, or food-preparation area. Stock your hygiene station with soap, paper towels, and hand sanitizer at the nearest sink.

- [ ] **Secure the room from pets.** Latch the door to the brooder room before bringing the chicks inside. Keep dogs and cats out to prevent panic from shadows, barking, or sudden movement.
- [ ] **Provide room-temperature water.** Fill the chick waterers just before leaving the house. Cold water can chill newly arrived chicks, lowering their core temperature to dangerous levels.
- [ ] **Plan safe transport.** Use a climate-controlled vehicle cab, not a sealed trunk or an open truck bed. Secure the cardboard transport box so it cannot slide if you brake suddenly.
- [ ] **Conduct an arrival health and stress check.** When placing chicks into the brooder, you can gently tap their beaks against the water surface to show them where to drink. Step back and observe. Soft peeping and active exploration are normal. Shrill, continuous peeping indicates distress. Huddling tightly in a pile means they are too cold; pressing against the far edges of the brooder means they are too hot.

Pathway 2: The pullet arrival checklist

Pullets are fully feathered young birds moving straight into an outdoor coop. Your primary focus is predator-resistant containment and minimizing the stress of transitioning to a new environment.

- [ ] **Inspect the predator envelope.** Walk the coop and run perimeter to verify every latch is closed, all hardware cloth is fastened securely with heavy-duty staples or screws, and no digging gaps exist beneath the aprons.
- [ ] **Add fresh, deep bedding.** Spread a generous layer of clean pine shavings or your chosen bedding on the coop floor and inside the nest boxes.
- [ ] **Prepare transport crates.** Use well-ventilated, secure crates or heavy-duty cardboard boxes lined with grip-friendly material, like a towel or shavings, so the birds do not slide. Secure the crates in the vehicle cab or an open cargo area. Lay a tarp under the crates to catch droppings.
- [ ] **Bring a breathable cover.** Have a thin, breathable sheet or towel ready to drape over the transport crates. Darkness naturally calms panicked birds during the drive, but ensure the fabric does not trap heat or block airflow.
- [ ] **Position multiple water stations.** Place the primary feeder and waterer inside the secure run. Because pullets often hide in corners when they first arrive, place an extra, stable, shallow water dish near their likely hiding spots to encourage hydration.
- [ ] **Enforce the quiet-settling rule.** Do not invite neighbors, extended family, or friends over to view the flock on day one. Open the transport crates inside the closed run, step back, and allow the pullets to walk out on their own timeline.
- [ ] **Observe from a distance.** Cautious pacing or freezing in place is normal for new pullets. Watch their breathing from afar. Do not chase them to conduct physical health checks on the first day unless you spot active bleeding or severe injury.

Pathway 3: The adult hen arrival checklist

Moving adult hens is highly disruptive to their established habits and social structures. Your focus must be on verifying their health history, executing strict biosecurity, and mitigating transition stress.

- [ ] **Verify health and source records.** Before handing over payment or loading the birds, ask for vaccination records, age estimates, and laying history. Confirm the specific brand and formulation of feed they currently eat so you can provide an exact match, preventing sudden digestive shock.
- [ ] **Execute an on-site health screen.** Observe the hens before they are boxed. Look for clear eyes, clean vents, smooth breathing, and active behavior. If a bird is lethargic, sneezing, or visibly unwell, do not bring it home. Walk away.
- [ ] **Plan your biosecurity boundaries.** Wear dedicated shoes to the seller's property. Do not wear those shoes into your house or your own chicken run. Wash or sanitize your hands immediately after loading the birds and before touching your steering wheel.
- [ ] **Set up strict quarantine.** If you are adding these hens to an existing flock, set up a strict quarantine zone separated from your main birds by as much distance as your property allows; thirty feet or more is a practical target when feasible, but follow your veterinarian, Extension specialist, or state diagnostic lab guidance for your situation. This area must be fully enclosed and stocked with its own dedicated feed, water, and footwear to prevent cross-contamination.
- [ ] **Prepare for an unhurried release.** Release adult hens directly into the enclosed run. Leave them completely alone to map their new surroundings without human interference.
- [ ] **Monitor for stress-induced illness.** Transport stress frequently causes underlying, hidden illnesses to surface in adult hens. Watch their droppings, breathing, and posture closely over the first week.

First observation notes

Once your birds are placed in the brooder or the secure run, step back and watch them for ten to fifteen minutes to establish a normal behavioral baseline. Keep a notebook near the coop or brooder to log your initial observations.

Do not try to diagnose illnesses on the first day. Your goal is simply to record their posture, breathing, and activity level. If you notice a bird standing apart from the flock with fluffed feathers, or struggling to breathe, document it and monitor closely for escalation.

Use this format to log your first observations:

- Date and time arrived: ________
- Number of birds: ________
- First water and feed interaction: ________
- Energy, breathing, droppings, and settling behavior: ________

Completed first-day observation example:

April 12, 2:00 PM. Four pullets arrived. All four birds walked out of the transport crate within ten minutes. They found the extra shallow water dish immediately. No sneezing, wheezing, limping, or lethargy noted. Paced the perimeter for twenty minutes, then settled quietly into the back corner of the run by 3:00 PM.

Appendix K. First 90 days care calendar

When you first bring chickens into a suburban backyard, the novelty makes every task feel important. You assume you will remember exactly how much feed they eat, when you last bought bedding, or what a healthy bird's breathing sounds like. But balancing a new flock with your existing household and work schedule will quickly crowd your memory.

This 90-day calendar bridges the gap between your flock's arrival and your long-term neighborhood operating plan. It tracks the daily, weekly, and monthly data you need to ensure your budget remains realistic, your neighbors remain unbothered, and your setup serves your lifestyle rather than consuming your weekends.

Core routine checklists

Before logging your weekly notes, establish your standard minimums. Use the weekly grids below to check off your daily visits and record your observations.

Daily baseline tasks:

- [ ] Check and refill water (ensure no freezing or algae growth).
- [ ] Check and top off feed (secure feed at night against rodents).
- [ ] Perform health scan (observe normal posture, activity, and breathing without handling).
- [ ] Collect eggs (if starting with laying-age hens).
- [ ] Secure the coop at dusk (verify all latches and inspect for predator gaps).

Weekly baseline tasks:

- [ ] Scrub waterers to remove biofilm.
- [ ] Manage manure (remove heavily soiled bedding or turn over litter).
- [ ] Audit hardware (check for loose wire, digging attempts, or compromised aprons).
- [ ] Log your budget and supply usage.

Month 1: Establishing the baseline (Days 1–30)

Your first thirty days are about settling the flock, measuring actual feed and water consumption, and identifying chore friction.

Week 1 (Days 1–7): Quiet arrival

- **Daily log:** [ ] Mon [ ] Tue [ ] Wed [ ] Thu [ ] Fri [ ] Sat [ ] Sun
- **Feed/Water notes:** How long does your primary waterer and feeder take to empty? Record your baseline refill rate here: ___________
- **Health/Bedding notes:** Describe the flock's baseline resting posture, breathing sounds, and manure output while they are healthy. ___________
- **Egg count:** ___________ (Note: 0 if raising chicks or young pullets)
- **Maintenance/Budget notes:** Complete initial neighbor check-in. Note any immediate chore friction, such as awkward latches or tripping hazards. ___________

Week 2 (Days 8–14): Chores and spills

- **Daily log:** [ ] Mon [ ] Tue [ ] Wed [ ] Thu [ ] Fri [ ] Sat [ ] Sun
- **Feed/Water notes:** Are birds scratching feed into the bedding? Adjust feeder height if necessary to minimize waste. ___________
- **Health/Bedding notes:** Are droppings accumulating rapidly under the roost? Determine your required cleaning frequency. ___________
- **Egg count:** ___________
- **Maintenance/Budget notes:** Fix the chore friction points you identified last week. ___________

Week 3 (Days 15–21): Odor and moisture

- **Daily log:** [ ] Mon [ ] Tue [ ] Wed [ ] Thu [ ] Fri [ ] Sat [ ] Sun
- **Feed/Water notes:** ___________
- **Health/Bedding notes:** Check for dampness around the waterer. Wet bedding creates ammonia and attracts flies. ___________
- **Egg count:** ___________
- **Maintenance/Budget notes:** Where is your soiled bedding or manure being stored? Verify your compost setup meets local waste rules. ___________

Week 4 (Days 22–28): Day 30 review

- **Daily log:** [ ] Mon [ ] Tue [ ] Wed [ ] Thu [ ] Fri [ ] Sat [ ] Sun
- **Feed/Water notes:** __________
- **Health/Bedding notes:** __________
- **Egg count (Month 1 Total):** __________
- **Maintenance/Budget notes:** Log all Month 1 expenses. Compare your actual feed and bedding costs to your initial estimates. Check in with neighbors regarding any noise or odor. __________

Month 2: Normalizing the routine (Days 31–60)

Now that the initial excitement has passed, Month 2 focuses on stabilizing your chores so they flow naturally with your work schedule.

Week 5 (Days 29–35): Time audit

- **Daily log:** [ ] Mon [ ] Tue [ ] Wed [ ] Thu [ ] Fri [ ] Sat [ ] Sun
- **Feed/Water notes:** __________
- **Health/Bedding notes:** __________
- **Egg count:** __________
- **Maintenance/Budget notes:** Time your morning and evening routines. If checking the birds takes more than ten minutes per visit, identify what is slowing you down. __________

Week 6 (Days 36–42): Fencing and predator audit

- **Daily log:** [ ] Mon [ ] Tue [ ] Wed [ ] Thu [ ] Fri [ ] Sat [ ] Sun
- **Feed/Water notes:** __________
- **Health/Bedding notes:** __________
- **Egg count:** __________
- **Maintenance/Budget notes:** Walk the run perimeter. Check hardware-cloth aprons for signs of digging. Inspect top coverage for gaps or sagging. __________

Week 7 (Days 43–49): Weather and environment

- **Daily log:** [ ] Mon [ ] Tue [ ] Wed [ ] Thu [ ] Fri [ ] Sat [ ] Sun
- **Feed/Water notes:** Note how weather changes affect daily water intake or freezing risks. __________
- **Health/Bedding notes:** Check coop ventilation. Is condensation or excessive dust building up inside? __________
- **Egg count:** __________
- **Maintenance/Budget notes:** __________

Week 8 (Days 50–56): Day 60 review

- **Daily log:** [ ] Mon [ ] Tue [ ] Wed [ ] Thu [ ] Fri [ ] Sat [ ] Sun
- **Feed/Water notes:** __________
- **Health/Bedding notes:** __________
- **Egg count (Month 2 Total):** __________
- **Maintenance/Budget notes:** Review your supply run rate. How many bags of feed and bales of bedding are you actually using per month? Update your ongoing budget. __________

Month 3: Stabilization and forward planning (Days 57–90)

By Month 3, you are no longer a beginner making guesses; you are a suburban flock manager making decisions based on real data.

Week 9 (Days 57–63): Hygiene and handling

- **Daily log:** [ ] Mon [ ] Tue [ ] Wed [ ] Thu [ ] Fri [ ] Sat [ ] Sun
- **Feed/Water notes:** __________
- **Health/Bedding notes:** Check your household boundary plan. Are dedicated coop shoes staying outside? Are handwashing rules being followed? __________
- **Egg count:** __________
- **Maintenance/Budget notes:** __________

Week 10 (Days 64–70): Pest and rodent check

- **Daily log:** [ ] Mon [ ] Tue [ ] Wed [ ] Thu [ ] Fri [ ] Sat [ ] Sun
- **Feed/Water notes:** Verify feed is securely locked up overnight. Sweep up spilled feed in the run. __________
- **Health/Bedding notes:** Inspect for flies or signs of rodents (droppings, chewed wood, or burrows). __________
- **Egg count:** __________
- **Maintenance/Budget notes:** Implement IPM-backed trapping or fly control if needed, ensuring strict label compliance. __________

Week 11 (Days 71–77): Flock dynamics

- **Daily log:** [ ] Mon [ ] Tue [ ] Wed [ ] Thu [ ] Fri [ ] Sat [ ] Sun
- **Feed/Water notes:** __________
- **Health/Bedding notes:** Observe the flock pecking order. Ensure subordinate birds have unhindered access to feed and water. __________
- **Egg count:** __________
- **Maintenance/Budget notes:** __________

Week 12 (Days 78–90): Day 90 review *(Note: Week 12 covers the final stretch to day 90).*

- **Daily log:** [ ] Mon [ ] Tue [ ] Wed [ ] Thu [ ] Fri [ ] Sat [ ] Sun
- **Feed/Water notes:** __________
- **Health/Bedding notes:** __________
- **Egg count (Month 3 Total):** __________
- **Maintenance/Budget notes:** Total your expenses for the first quarter. Use this data to project your annual flock cost. __________

Day 90 final operating plan review

After ninety days, consolidate your notes into a permanent routine. Sit down and answer these five questions to finalize your setup:

- Are the daily chores manageable, or do I need larger capacity feeders and waterers to get through the work week?
- Is my manure management system working, or is it creating odors, flies, or drainage problems?
- Are my neighbors still comfortable, and have I proactively followed up with them?
- Have I fully adhered to my local legal limits, zoning setbacks, and HOA guidelines?
- Are household hygiene rules, such as handwashing, dedicated shoes, and keeping poultry out of the kitchen, being strictly followed by all family members and guests?

Appendix L. Daily, weekly, monthly, and seasonal routine tracker

WHEN YOU FIRST BRING a small flock of laying hens into a suburban backyard, the novelty makes every task feel important. It is easy to assume you will naturally remember to check the water, grab the eggs, and lock the coop door. You might look at the idea of a printed tracker and assume a formal checklist is overkill for three or four hens.

But when routines remain casual in a busy suburban household, essential tasks slip. A busy Tuesday evening or a sudden change in your family schedule can easily distract you from the backyard. Consistency is your primary tool for noticing problems early and keeping your setup neighbor-friendly. Checking the exact same things in the exact same order every single day establishes a baseline. This helps you spot a slightly damp corner of the coop before it generates an ammonia odor, or a small gap in the hardware cloth apron before a raccoon finds it.

How to use the routine tracker

To get the most out of them, establish a few ground rules for your household:

- **Keep it visible:** Print the pages, attach them to a clipboard, and hang the tracker near your back door, inside the garage, or near your dedicated egg-washing station. Visibility ensures accountability.
- **Assign initials:** Write your initials in the box when you complete a task. When multiple family members share the workload, a physical tracker eliminates the guesswork of assuming someone else locked the coop.
- **Stay accurate:** Initial tasks only after they are completed. Do not check things off in advance.
- **Use the notes section:** Jot down anything unusual. If a hen is resting more than normal, a latch feels sticky, or the oyster shell bag is getting low, write it down immediately.

The daily baseline tasks

Your daily routine is about immediate health, hygiene, and security. It should take just a few minutes.

Feed and water check: Verify that the feeders are flowing and the water is clean. Ensure there is no algae buildup in the summer or ice accumulation in the winter. Chickens should not be left without clean water.

Egg collection: Gather eggs at least once a day, and preferably twice. Prompt collection helps prevent eggs from freezing in the winter, spoiling in the summer heat, or getting stepped on. It also reduces the chance that hens will develop the habit of eating their own broken eggs.

Bird observation: Take one minute to watch the flock before you interact with them. Are they all active? Are their eyes clear? Is anyone isolating in a corner or breathing heavily? Catching illness early relies on knowing what normal behavior looks like.

Spot cleaning: Scrape the droppings board or remove heavy manure accumulation under the roosts. In a suburban setting, daily spot cleaning is one of the most effective ways to prevent odors and fly blooms from bothering your neighbors.

Evening latch check: Once the hens have gone inside to roost at dusk, physically pull the coop door latch. Do not just look at it from the kitchen window. Touch the hardware to ensure it is locked tight against nighttime predators.

The weekly reset tasks

Pick one day a week, often a Saturday or Sunday morning, to perform a slightly deeper reset.

Scrub waterers: Empty the waterers completely and scrub them with a dedicated brush to remove the slippery biofilm that builds up on the inside.

Refresh nests: Remove any soiled bedding or accidental manure from the nesting boxes. Adding a fresh handful of pine shavings encourages the hens to keep laying there and ensures your eggs come out clean.

Rake the run: Turn over the mulch, sand, or wood chips in your outdoor run. This manages the manure load, promotes drainage, and prevents the ground from becoming a compacted, foul-smelling surface.

Check feed storage: Verify that your bulk feed is secure. Suburban rodents are opportunistic. Check your galvanized metal bins to ensure lids are sealing tightly and no spilled feed is lingering on the ground to attract mice.

The monthly inspection

Once a month, perform a physical inspection of your coop's structural integrity and check your long-term supplies. This prevents emergency trips to the farm store and keeps the shelter secure.

Hardware cloth and fencing: Push against the mesh. Check for loose staples, rusting wire, or pulled seams. **Latches and hinges:** Wood swells and shifts with the weather. Verify that all doors still close tightly without leaving gaps that a raccoon or weasel could pry open. **Perimeter check:** Walk the outside of the coop and run. Look for signs of attempted digging along the hardware-cloth apron. Check the roof for loose shingles or moisture leaks. **Inventory check:** Verify your levels of layer feed, grit, oyster shell, and pine shavings.

The seasonal review

Perform this broader review as the weather shifts.

Ventilation and weather: Ensure your high ventilation remains open but draft-free. Adapt your setup for summer shade requirements or winter wind blocks.

Run conditions: Address seasonal changes in the yard. Manage autumn leaf buildup, prepare for spring mud, or adjust for summer dust.

Life stage adjustments: Note seasonal changes in the flock. Expect egg production to drop or pause entirely during the shorter days of winter or during the fall molt.

Return to this system whenever your routines slip, a new caretaker takes over, or the seasons change. Keeping this tool front and center will help you maintain accountability across your household and ensure your suburban flock stays healthy, contained, and quiet.

Appendix M. Egg handling and household hygiene checklist

When you see your backyard hens looking bright and active every day, it is easy to assume, "Our birds are healthy, so we do not need to worry about germs." But healthy-looking poultry naturally carry bacteria such as Salmonella and Campylobacter in their droppings, on their feathers, and on their feet. Caution is not a sign that your flock is unusually dirty; it is standard public health practice.

This checklist establishes one conservative, U.S.-focused default for household hygiene and egg safety. Post a copy near your backdoor or mudroom drop zone to keep yard items outside, protect your kitchen, and safely handle fresh eggs.

Household Boundaries and The Drop Zone

Hygiene fails when it is inconvenient. Set up a clear physical boundary between the chicken yard and your living space before you bring birds home.

[ ] **Create a designated drop zone.** Pick a spot in the mudroom, the garage, or directly inside the back door you use to access the coop. Store your dedicated coop shoes and egg-collection basket here.

[ ] **Wear dedicated coop shoes.** Designate one pair of boots or slip-on shoes exclusively for the chicken yard. These will step in droppings and mud. Never wear them inside the living areas of your house.

[ ] **Keep poultry out of the house.** Chickens belong outside as the conservative household default. Keep birds out of kitchens, household bathrooms, and routine human living areas, not for a quick photo, not as a casual cold-night exception, and not to show a guest. If a bird requires medical isolation, use a dedicated crate in a garage or shed when appropriate.

[ ] **Keep coop equipment out of household sinks and tubs.** Feeders, waterers, scrub brushes, and scrapers should not be washed in the kitchen sink, bathroom sink, or bathtub. Wash all flock equipment outdoors using a garden hose or outdoor spigot. Store these cleaning tools outside.

Personal Hygiene and Vulnerable People

[ ] **Wash hands with soap and water.** Wash your hands immediately after every flock visit, whether you touched a hen, gathered an egg, or just opened the coop door.

[ ] **Understand hand sanitizer limits.** Hand sanitizer does not work effectively if your hands are visibly soiled with dirt, mud, or manure. It cannot penetrate organic matter. Use soap and warm water as your primary method, reserving sanitizer only as a temporary backup until you can reach a sink.

[ ] **Supervise children at the sink.** Children are the most likely to skip proper handwashing or to touch their faces after handling birds. Watch them wash to ensure they use soap and scrub for a full twenty seconds.

[ ] **Protect vulnerable household members.** Children under five, adults over sixty-five, pregnant people, and anyone with a weakened immune system are at a higher risk for severe illness from poultry-related bacteria. They should enjoy watching the flock from a safe distance and avoid direct contact with the birds, the coop, and unwashed eggs. (They can still safely consume properly handled, fully cooked eggs.)

Egg Collection and Sorting

[ ] **Use a dedicated, washable collection container.** Do not gather eggs in your pockets, your shirt hem, or a random kitchen bowl. Use a washable plastic, wire, or silicone basket meant only for daily gathering. If an egg cracks or soils the bottom, wash and sanitize the basket before using it again.

[ ] **Collect eggs daily.** Gather eggs at least once a day, or twice a day during extreme heat or freezing weather, to prevent spoilage, cracked shells, and dirty eggs.

[ ] **Discard cracked eggs immediately.** Bacteria easily penetrate a cracked shell. Do not attempt to salvage, wash, cook, or feed cracked eggs back to your flock. Discard them safely where pets, poultry, and wildlife cannot reach them.

[ ] **Discard heavily soiled eggs.** If an egg is caked in mud, yolk, or manure, throw it away. Do not try to scrub heavily contaminated eggs clean, as aggressive scrubbing can push bacteria through the porous shell.

The Conservative Default: Washing and Refrigeration

While unwashed eggs possess a natural protective coating called the bloom, keeping unwashed eggs on the kitchen counter introduces unnecessary risk for suburban beginners. This checklist uses the conservative public-health default.

[ ] **Wash eggs properly.** Do not soak eggs in a bowl of water, which can pull bacteria inside the shell. Instead, wash eggs under gently running water that is warmer than the egg itself (between 90°F and 120°F [VERIFY]). The warmer water makes the egg contents expand slightly, pushing dirt and bacteria outward.

[ ] **Use a designated egg scrubber.** If you use a brush or sponge to gently remove light debris, use it exclusively for washing eggs. Do not use your regular dish sponge.

[ ] **Dry eggs completely.** Immediately dry the washed eggs with a clean paper towel.

[ ] **Refrigerate immediately.** Place the clean, dry eggs into a carton and store them in the main body of the refrigerator, where the temperature remains a consistent 40°F or below. Do not store them in the refrigerator door, where temperatures fluctuate every time the appliance is opened.

Egg Gifting, Sharing, and Selling

Giving away extra eggs is a highlight of backyard chicken keeping, but it requires strict attention to safety, local laws, and carton labeling.

[ ] **Keep the oldest eggs; give away the freshest.** When gifting eggs, give away the most recently laid, freshly washed, and refrigerated ones to maximize shelf life for the recipient.

[ ] **Never share or sell cracked or dirty eggs.** Only share eggs that have intact shells and have been cleaned and refrigerated according to the conservative household default.

[ ] **Do not reuse commercial cartons without obscuring labels.** If local rules allow you to reuse grocery store cartons, completely cross out or cover any original farm name, USDA grade, expiration date, and nutritional claims so the carton is not misleading. Better yet, purchase blank, unused cartons for gifting and sharing.

[ ] **Verify local egg-selling rules.** If you plan to sell eggs, you must verify your state's cottage food laws, agricultural department regulations, and local zoning rules. Many jurisdictions strictly regulate carton labeling, washing protocols, refrigeration requirements, and whether you can sell from your home versus a farmer's market.

Neighborhood Visitors and Guest Hygiene

[ ] **Explain the rules before opening the gate.** Visitors will naturally want to touch the hens. State your household boundary rules plainly before anyone enters the yard.

[ ] **Give specific instructions to children.** Remind visiting children not to kiss the birds, snuggle them close to their faces, or put their fingers in their mouths.

[ ] **Guide guests to the sink.** Do not wait for a guest to ask where the sink is. When the coop visit is over, lead them directly to the handwashing station. Making this an automatic part of the backyard tour removes any awkwardness and ensures everyone stays safe.

Appendix N. Child, dog, and guest boundary plan

When you bring backyard chickens home, you might assume safety boundaries are obvious. But common sense varies widely depending on the species and the age of the individual. A toddler's common sense says to grab a squawking bird by the tail. A dog's common sense says to chase a fluttering object. A visiting neighbor's common sense might say to feed your hens potato chips through the wire.

Safety and hygiene boundaries must be strictly managed to prevent accidents, injuries, and illness before they happen. Complete this plan and establish your household rules before you bring your first birds home.

The dog separation plan

There are no safety guarantees with dogs and chickens. A gentle, older family dog who loves children and sleeps on the couch may still possess a deeply ingrained prey drive that activates the second a chicken flutters its wings.

Obedience training is not a reliable substitute for physical separation. A highly trained dog can fail in a moment of high excitement, and a single mistake is generally fatal for the chicken. To protect your flock and keep your dog out of stressful situations, rely on impenetrable barriers and strict access routines.

- **No guarantees:** Never assume your dog is "chicken-safe." Do not leave dogs and chickens unattended together, even if they have ignored each other for months.
- **Physical barriers:** Build the plan around physical separation whenever dogs and chickens could access the same space. Use sturdy, predator-resistant materials such as heavy-gauge hardware cloth for the run rather than flimsy chicken wire that a determined dog may be able to damage.
- **The shifting access rule:** If the dog is loose in the backyard, the chickens must be locked securely inside their run. If the chickens are allowed to free-range on the lawn under your supervision, the dog must be locked inside the house.
- **No fence-fighting:** Do not let dogs press their noses against the run wire to bark, dig, or pace. This causes unnecessary stress for the flock. Keep the dog away from the immediate perimeter of the run.

Child access and age-appropriate chores

Children and chickens require strict adult supervision. The Centers for Disease Control and Prevention advises that live poultry can carry Salmonella and Campylobacter in their droppings and on their feathers, even if the birds look perfectly healthy. For this reason, the CDC recommends that children under five years old should not handle live poultry, as their immune systems are still developing and they are more likely to put their hands in their mouths.

Enforce the strictest hygiene rule immediately: no kissing, no snuggling, and no holding birds near the face, ever.

Assign chores based on maturity, calmness, and physical strength. Young children must not pick up or carry live chickens. If a child loses their grip or a bird panics and flaps, the child will naturally drop the bird. Falling from a child's arms can cause a chicken to suffer broken legs, a fractured keel bone, or severe internal injuries. Wait until children are older and stronger.

- **Toddlers and children under five:** Do not touch the birds. Chores are limited to watching quietly from outside the run, carrying an empty egg basket, or tossing treats on the ground while holding an adult's hand.
- **Elementary-age children:** May scoop feed into a bucket under supervision, collect eggs using two hands, and help shut the coop door.
- **Older children and teens:** May refill heavy waterers, handle bedding, and double-check gate latches.
- **Adults only:** Lift, carry, or inspect live birds, administer first aid, open or close run gates, clean the coop, and manage the dog.

Guest rules card

Many friends, extended family members, and neighbors will want to see the backyard chickens. Left unguided, guests may poke their fingers through the wire, try to open the nesting boxes, or chase a hen to pick it up. You must manage their excitement and the interaction.

Set expectations before anyone steps into the backyard. Keep it friendly, but make clear that guests do not enter the run or feed anything to the birds unless provided by you. Write these rules on a card and keep it near your back door.

Our Backyard Flock Rules

- **Do not open gates:** We manage the doors to keep the birds safe and contained.
- **Look, don't touch:** Please keep fingers out of the wire. Chickens explore the world with their beaks and may peck at painted fingernails or rings.
- **No chasing:** Chickens are prey animals. Chasing them causes extreme stress.
- **Approved treats only:** We will provide you with safe treats, like dried mealworms, to toss on the ground. Please do not feed them human food.
- **Wash up:** Everyone must wash their hands with soap and water after visiting the coop area.

Garden zones and boundaries

Chickens are highly destructive foragers. If given access to vegetable beds or prized flower borders, they will quickly scratch up seedlings, eat ripe tomatoes, take dust baths in freshly turned soil, and scatter mulch across the lawn.

Furthermore, raw poultry manure poses a significant food-safety risk if deposited in edible garden beds. University Extension guidelines strongly advise against allowing chickens to roam through beds where root vegetables or leafy greens are grown for human consumption.

If a space is not physically fenced off, it belongs to the chickens. Use temporary step-in fencing, hardware-cloth cloches, or raised bed covers. Outline your zones:

- **Chicken-approved areas:** (e.g., specific compost pile, fenced lower lawn)
- **Off-limits garden beds:** (e.g., raised vegetable beds, front porch, patio)
- **Barriers used:** (e.g., wire cloches over seedlings, latched garden gate)

Household hygiene and footwear station

Hygiene fails when it is inconvenient. Set up a clear physical boundary between the chicken yard and your living space.

- **Coop shoes:** Designate a specific pair of boots or slip-on shoes for the chicken yard. Leave them outside the back door or in a mudroom tray. Do not wear coop shoes into the kitchen. This prevents you from tracking microscopic poultry waste across the floors where toddlers crawl or pets sleep.
- **Handwashing:** Wash your hands thoroughly with soap and warm water immediately after touching birds, eggs, bedding, feeders, waterers, soil, or coop tools. Hand sanitizer is a temporary backup; it does not replace mechanical scrubbing at the sink, especially when organic matter like dirt or manure is present on your hands.

Boundary breach response

Even the best boundary plans occasionally fail. A gate latch slips, a dog digs under a fence, or a guest leaves the run door open. If a boundary breaks, do not yell, as loud noises will further scatter a panicked flock. Follow this response protocol:

- **Secure the threat first:** If a dog breaches the yard or run, immediately leash and remove the dog and secure it inside the house before doing anything else.
- **Remove children and guests:** Get vulnerable humans calmly out of the run area and away from the chaos. Wash any minor scratches with soap and water.
- **Close gates:** Contain the remaining flock so no more birds escape.
- **Count the birds:** Ensure everyone is accounted for.
- **Check for obvious injuries:** Evaluate the flock calmly once the yard is quiet.
- **Seek professional help if needed:** For serious wounds, use your emergency contact sheet to reach an avian veterinarian or animal control. Do not attempt complex veterinary surgery yourself.
- **Review the failure:** Identify exactly what boundary failed (a loose latch, a dug-out perimeter, an unlatched gate) and fix it before allowing yard access again.

Appendix O. Sick-bird observation log and help-call worksheet

When a backyard chicken looks sick, your immediate job is not to become a veterinarian. Searching the internet for symptoms usually leads to misdiagnosis, and blindly applying a remedy you read about online can do more harm than good. Instead, your job is to observe, isolate or contain the bird as advised, document the facts, and call for qualified help.

Experts, such as an avian veterinarian, your county Extension office, a state poultry diagnostic lab, or the state veterinarian, cannot provide help without accurate information. Vague descriptions like "my hen looks sad and will not eat" apply to almost every poultry ailment. Taking a few minutes to write down exactly what you see gets your bird the right help.

Triage and safety warnings

Before you fill out this log, pause and review these three critical safety rules:

- **Do not transport sick or dead birds unless instructed.** If you suspect a highly contagious disease, if multiple birds are sick at the same time, or if you have experienced sudden unexplained flock deaths, keep the birds on your property. Transporting them risks spreading disease across county lines and into veterinary clinic waiting rooms. Call your state veterinarian, local diagnostic lab, or Extension office first, and follow their exact containment and transport instructions.
- **Do not blindly medicate.** Do not administer leftover antibiotics, random farm-store dewormers, pain medications, or internet home remedies without professional guidance. Many medications are strictly regulated or prohibited in laying hens. Even approved medications require strict, prolonged egg-withdrawal periods, meaning you cannot safely eat, share, or sell the eggs for a set number of weeks.
- **Protect household hygiene.** Wash your hands thoroughly with soap and warm water immediately after handling a sick bird or inspecting droppings. Dedicate specific footwear and clothing for the sick-bird area, and keep children, dogs, and house guests completely away from the affected bird and its enclosure.

How to gather the right clues

Before you call an expert, stand back and watch your flock. Chickens instinctively hide weakness to avoid predators, so observe the affected bird when she does not know you are looking. Gather specific details for the following categories:

- **Flock context:** Experts need to know if this is an isolated incident. Are other birds showing similar signs? Have you introduced new birds recently, or have wild birds been accessing your feeders?
- **Posture and movement:** Notice if her tail feathers are pointed down toward the ground or if her head is tucked tightly into her shoulders. Is she limping or hiding?
- **Breathing:** Most healthy chickens breathe quietly. Listen closely for a clicking sound, a wheeze, or a wet rattle. Note if she is holding her beak open to pull in air, or if her tail is bobbing forcefully with every breath.
- **Droppings:** Look under the roosting bar where the sick bird slept the night before. Note the exact color and consistency.
- **Appetite and water:** Pay attention to whether she is eating or drinking normally, consuming less than usual, or actively refusing her favorite foods.
- **Photos and video:** Take a clear, well-lit photo of the sick bird's droppings. Next, record a ten-second video of the bird standing or resting. A short video shows an expert her breathing rate, posture, and subtle movements far better than a still photograph.

Observation log

Do not guess a disease name unless a qualified professional has given you one. Stick entirely to observable facts. Follow this exact structure to gather information before you make your call.

Flock and Household Context

- **Flock size:** (Total number of birds)
- **Birds affected:** (How many are showing signs?)
- **Deaths:** (Have any birds died suddenly? If yes, how many and when?)
- **New birds:** (Have you added any new chicks, pullets, or adults in the last thirty days?)
- **Wild-bird exposure:** (Have wild birds been in the run or accessing the feed and water?)

Bird Profile

- **Bird ID:** (Name, breed, age, and approximate weight if known)
- **Egg status:** (When did she last lay an egg? Was the shell normal, thin, or missing?)

Observation Log

- **Date and time noticed:** (When you first spotted the problem)
- **Symptoms:** (What specifically looks or sounds different about this bird compared to the rest of the flock?)
- **Movement and posture:** (Is she limping, drooping her wings, hiding in the corner, or keeping her eyes closed?)
- **Breathing:** (Describe any noises, open-beak breathing, tail bobbing, or coughing)
- **Appetite and water:** (Is she eating feed normally? Consuming less than usual? Refusing favorite treats? Is she drinking water?)
- **Droppings:** (Describe the exact color, frequency, and consistency. Are they solid, liquid, or foamy?)
- **Feed/water changes:** (Have you recently introduced a new brand of feed, supplements, or a new water source?)
- **Treatments:** (List anything you have already given the bird or applied to the coop, even if just vitamins or electrolytes)

Media Checklist [ ] Clear photo of droppings taken [ ] Ten-second video of the bird standing or resting taken

Help-Call and Follow-Up

- **Contacts:** (Name and phone number of the vet, Extension office, diagnostic lab, or state veterinarian)
- **Advice received:** (Exactly what the expert instructed you to do right now, including isolation and transport instructions)
- **Follow-up:** (What practical steps you actually took based on the professional advice)

Completed sample log (example)

Notice how this keeper sticks entirely to observable facts without trying to guess the diagnosis. Do not infer a medical condition from this example; use it simply as a model for how to record clear, factual observations using short notes.

Flock and Household Context

- **Flock size:** 6 hens.
- **Birds affected:** 1.
- **Deaths:** 0.
- **New birds:** None added in the past year.
- **Wild-bird exposure:** Yes, wild sparrows frequently get into the run and eat from the feeder.

Bird Profile

- **Bird ID:** Hazel, Speckled Sussex, 18 months old.
- **Egg status:** Did not lay yesterday or today. Last egg was collected Sunday and had a perfectly normal shell.

Observation Log

- **Date and time noticed:** Tuesday, October 12, 8:00 AM.
- **Symptoms:** Standing by the fence away from the rest of the flock. Eyes closed most of the time. Feathers around the vent area are heavily soiled.
- **Movement and posture:** Tail feathers are pointed straight down toward the ground. Neck is pulled in tight to her body. Walks very slowly when moved, and immediately stops to tuck her head back in.
- **Breathing:** Quiet. Panting slightly with beak open even though it is only sixty degrees outside. No clicking, wheezing, or rattling noises heard.
- **Appetite and water:** Refused scrambled eggs, which are usually her favorite treat.
- **Droppings:** Checked under her specific spot on the roost. Droppings are liquid, mostly clear and white.
- **Feed/water changes:** Switched to a new brand of layer feed last Thursday. Rained heavily all weekend.
- **Treatments:** None given. Have not administered any medications.

Media Checklist [x] Clear photo of droppings taken [x] Ten-second video of the bird standing or resting taken

Help-Call and Follow-Up

- **Contacts:** Dr. Evans at Maple Veterinary Clinic.
- **Advice received:** Based on the information provided, one affected bird and no sudden flock deaths, the clinic advised that Hazel could be transported for the 2:00 PM appointment if she was contained and handled with hygiene precautions.
- **Follow-up:** Set up the wire dog crate in the attached garage, moved Hazel, loaded the pet carrier into the car for the appointment, and thoroughly scrubbed hands and boots with hot, soapy water.

Appendix P. Neighbor conversation scripts and complaint response plan

Proactive communication replaces neighbor assumptions with a face, a name, and a sensible plan. If you build a coop quietly, unexpected structures or new sounds can trigger complaints to local authorities before you even know an issue exists. Initiating a conversation early gives neighbors a direct, calm path to reach you.

However, keeping a backyard flock involves strict legal and neighborhood realities. You must handle neighbor interactions carefully. Your goal is to maintain neighborhood peace without admitting legal liability, making impossible guarantees, or inviting frivolous damage claims.

Use the following scripts as starting points. Customize them to fit your speaking style, but adhere to four strict boundaries:

- **Do not admit liability:** You can apologize for a disruption without claiming legal or financial fault.
- **Do not overpromise:** Never guarantee a "zero-odor" or "completely silent" flock. Promise a well-managed setup.
- **Do not solicit damage claims:** Never ask a neighbor if a loose bird ruined their landscaping or offer preemptive compensation.
- **Do not record conversations:** Do not attempt to secretly audio-record your neighbors. This can violate state recording laws and instantly destroy trust. Document interactions using a written log instead.

Before the flock arrives

Catch your neighbor while they are working in the yard before you buy materials or bring a single bird home. Keep it brief.

"Hi, I wanted to give you a quick heads-up about a project we are setting up in our backyard. We have verified the local guidelines, and we are going to keep a small flock of laying hens, absolutely no roosters. I know people sometimes worry about noise or mess, so I wanted to let you know we are putting a lot of work into a clean, well-managed setup. If you ever have any questions or concerns once they arrive, please let me know directly so I can address them."

Why this works: It establishes that you know the local laws, confirms there will be no roosters, and invites direct communication rather than a premature call to animal control.

General concern response

Even with perfect planning, a neighbor might eventually knock on your door with a complaint. Fight the urge to get defensive. Your immediate job is to de-escalate the situation and buy yourself time to investigate.

"Thank you for letting me know. I want to make sure our setup is not causing a disruption. I am going to head back into my yard right now to take a close look at what is going on, and I will see what adjustments I can make."

Why this works: You validate their decision to speak with you without agreeing that a nuisance currently exists. You commit only to inspecting your own property.

Odor concern response

Odor is a common concern. Take it seriously, but do not confess that your coop is a filthy nuisance.

"I appreciate you bringing this up. A properly managed run should not be disruptive at the property line. I am going to inspect the coop bedding right away. If the weather or a spill has created excess moisture, I will turn or replace the bedding today to get it drying out. I will check back with you in a couple of days to see if you notice an improvement."

Why this works: It acknowledges the standard without admitting fault. It focuses entirely on the mechanical, corrective action of managing moisture.

Noise concern response

If a neighbor complains about noise, you need to find out exactly when it happens so you can manage the trigger.

"Thank you for telling me. Hens are generally fairly quiet, but occasionally something specific in the environment sets them off. Can you tell me what time of day you usually notice this? I am going to monitor our yard closely over the next few days during those times to identify the trigger, and I will see what I can do to manage the disruption."

Why this works: Noise issues are often tied to specific triggers, such as a delivery truck or a stray cat on the fence. Asking for times shifts the conversation from a debate to a factual observation.

Boundary and loose hen response

No matter how carefully you build your run, a curious hen might occasionally slip through a gap. Go to their door promptly.

"I apologize for the disturbance today. One of the hens crossed the property boundary. I have already secured her back inside our run. I will be inspecting our fence line and run enclosures this evening to secure the boundary and prevent this from happening again."

Why this works: It addresses the event directly and outlines your corrective action. Crucially, it does not ask, "Did she ruin your garden?" Never solicit damage claims. Furthermore, if a bird is currently in a neighbor's yard, ask for clear permission before crossing the property line to retrieve it, and do not suggest that you are entitled to enter without consent.

Complaint response log template

Handling the immediate conversation is only half the job. To protect your flock and document a pattern of responsible care, you must log neighbor concerns and your actions. Do not rely on your memory. Keep blank copies of this template in your flock record binder.

- Date and time:
- Neighbor contact name:
- Nature of the complaint:
- Inspection findings:
- Corrective action taken:
- Follow-up date:
- Follow-up result:

Completed response log example)

Here is what a professional, objective entry looks like in practice.

- **Date and time:** August 14, 10:00 a.m.
- **Neighbor contact name:** Mark (south fence line).
- **Nature of the complaint:** Mark stated there were an unusual number of flies near his patio and asked if they were coming from our yard.
- **Inspection findings:** Walked the run. Found a damp patch of straw under the primary waterer.
- **Corrective action taken:** Removed the wet bedding, leveled the ground to prevent pooling, and replaced the straw with coarse construction sand to improve drainage. Hung two disposable fly traps on the far side of the run.
- **Follow-up date:** August 17.
- **Follow-up result:** Spoke with Mark at the fence. Asked if the fly situation had improved. He stated the patio was much better and thanked me.

Follow-up checklist

Following up proves you are a reliable neighbor. Keep this sequence in mind for every interaction:

1. **Listen carefully.** Acknowledge the concern without interrupting or debating local ordinances.
2. **Inspect promptly.** Look closely at the coop, run, or fencing on your own property as soon as the conversation ends.
3. **Take corrective action.** Clean the bedding, secure the physical boundary, or block the noise trigger.
4. **Document the event.** Fill out your complaint response log immediately while the facts are fresh.
5. **Follow up.** Check back within 48 to 72 hours to confirm the adjustment made a difference.

Appendix Q. Completed sample first-flock operating plan

THIS FICTIONAL EXAMPLE DEMONSTRATES the level of detail a finished suburban chicken plan requires. While the specific municipal laws, homeowner's association rules, and property details in this sample are entirely made up, the planning process shows how all the book's decisions connect. Translating abstract rules into a documented, highly specific plan is what prevents nuisance complaints, unexpected expenses, and panic when problems arise.

Use this sample as a benchmark. Compare your own notes against these decisions so you know what a practical, written operating plan looks like before you commit to buying birds.

Sample household parameters

- **Owner:** Fictional beginner named Abigail.
- **Property:** Suburban lot with limited shaded space and thick tree roots near the ideal coop location.
- **Household limits:** One child (age six) and one high-energy dog (Golden Retriever).

Legal status and neighbor plan

- **City zoning verification:** Checked via the city ordinance portal on March 1st. The fictional code allows up to six hens, strictly prohibits roosters, and requires coops to be kept at least fifteen feet from any neighboring property line.
- **HOA verification:** Checked the neighborhood portal. No specific poultry ban exists, but the bylaws require all accessory outbuildings, including chicken coops, to be hidden from street view behind a six-foot privacy fence.
- **Neighbor outreach:** Spoke directly with the neighbors on both sides of the fence before buying any supplies. Shared the placement plan, promised there would be no roosters, and provided a direct cell phone number.
- **Response plan:** If a neighbor calls or texts about noise, flies, or odors, the policy is to respond the same day. Abigail will immediately check the coop bedding, remove any wet spots, and adjust the cleaning schedule to eliminate the issue.

Flock and space decisions

- **Flock size:** Three hens. Even though the city allows six, starting with three prevents overcrowding, manages neighbor impact, and keeps waste management simple in a small yard.
- **Space verification:** Three large hens require a conservative minimum of 12 square feet of interior coop space (4 square feet per bird) and 30 square feet of enclosed run space (10 square feet per bird). Abigail selected a sturdy structure measuring 16 square feet inside and 40 square feet in the run, easily accommodating three birds without maxing out capacity.
- **Source and age:** Ordered three twelve-week-old started pullets through an NPIP-certified local farm supplier. This bypasses the indoor brooder phase, avoids the electrical fire risks of indoor heat plates, and minimizes the risk of accidentally bringing home a surprise rooster.
- **Breed:** Buff Orpingtons. Selected for their cold-hardy traits, confinement tolerance, and generally calm temperament in neighborhood settings.

Coop placement, predators, and waste

- **Location:** Behind the detached garage. This spot meets the fifteen-foot legal setback, hides the structure from the street to meet HOA rules, provides a windbreak during winter storms, and receives dappled shade from a mature oak tree.
- **Utility locate:** Contacted the 811 utility-locate system before finalizing the placement or attempting to dig trenches, then checked whether any private owner-installed lines needed separate locating.
- **Predator upgrades:** Replaced the prefabricated coop's flimsy wooden toggles with heavy-duty carabiner latches to deter raccoons.
- **Apron workaround:** Thick oak tree roots blocked the planned twelve-inch trench for a buried hardware-cloth skirt. Instead, Abigail laid a twenty-four-inch hardware-cloth apron flat on the ground extending outward from the run. She secured it tightly with landscaping staples and heavy stone pavers to deter digging predators.
- **Waste management:** Verified that municipal codes allow enclosed composting. Placed a heavy-duty, rodent-resistant dual-chamber compost tumbler twenty feet from the property line. All soiled pine shavings and droppings go into the tumbler to cure, strictly following local integrated pest management (IPM) guidelines to prevent flies and rats.

Budget and supplies

- **Startup costs ($1,200 to $1,600):** This covers the coop and run, rolls of half-inch hardware cloth, heavy-duty latches, three NPIP-certified pullets, a galvanized steel feed bin with a tight-fitting lid, an outdoor-rated waterer, a hanging feeder, and dedicated slip-on boots.
- **Monthly ongoing costs ($45 to $65):** Covers high-quality pelleted layer feed, fresh pine shavings, oyster shell, and minor first-aid replenishment.
- **First-year total estimate ($1,740 to $2,380):** Covers the high-end startup costs plus twelve months of the highest ongoing expenses. This reflects the reality that backyard eggs are not a money-saving venture.

Household boundaries and egg hygiene

- **Dog boundaries:** The dog is completely fenced away from the coop area. Because no dog can ever be guaranteed safe around poultry, the dog is never allowed in the yard at the same time the chickens are out of their run.
- **Child boundaries:** The six-year-old is taught not to kiss the chickens, touch her face after holding them, or carry them inside the house.
- **Handwashing protocol:** Anyone who touches the birds, eggs, or coop hardware must immediately wash their hands with soap and water at the garage utility sink before touching anything else.
- **Conservative egg handling:** Collect eggs daily. Discard any cracked, leaking, or heavily soiled eggs. Immediately refrigerate intact, visually clean eggs unwashed. Wash eggs under warm running water only immediately before cooking.

Daily and weekly routines

- **Morning (5 minutes):** Put on dedicated coop boots. Unlock the carabiners, check the waterer, top off the feed, and do a quick visual scan to ensure no bird is acting lethargic, limping, or hiding.
- **Evening (5 minutes):** Do a physical headcount at dusk, securely lock all heavy-duty latches, and collect eggs in a dedicated plastic basket.
- **Weekly (20 minutes):** Scrape the dropping boards, add a fresh layer of pine shavings to the coop floor, and scrub and rinse the waterer to prevent biofilm buildup.

Health observation and emergency escalation

- **Information access:** Emergency contact sheet is tacked inside the garage near the feed bin. It lists a local avian veterinarian, the county Extension office, and the state agricultural diagnostic lab.
- **Sick-bird protocol:** If a bird shows signs of illness, immediately isolate her in a dedicated dog crate inside the garage. Observe her condition. Do not prescribe, dose, or attempt home remedies.
- **Transport rules:** If sudden death occurs or a contagious disease is suspected, do not transport sick or dead birds to a local clinic unless explicitly instructed to do so by the state veterinarian or diagnostic lab.
- **Backup care:** A trusted coworker who already keeps a biosecure flock has agreed to be an emergency vacation sitter.

Seasonal weather preparations

- **Summer heat:** Monitor the flock for heat stress and ensure the oak tree's dappled shade provides enough relief during peak afternoon sun. Provide extra water stations.
- **Winter freeze:** Install an outdoor-rated heated waterer base before the first hard freeze. Power it using only a manufacturer-approved, heavy-gauge outdoor cord plugged directly into a GFCI-protected, weather-covered exterior outlet. Keep cords completely out of the birds' reach.

By writing these specifics down, Abigail removed the guesswork from her first ninety days. She knows her municipal limits, she mapped out her daily steps, and she knows exactly who to call if an illness or emergency arises. Follow this rigorous model to build your own readiness plan.

Appendix R. Trusted resources for current rules, health, biosecurity, and food safety

Because rules, zoning overlays, and public-health guidance change frequently, establishing a directory of primary sources before you buy birds is the only way to protect your flock and your household. Below is the breakdown of the exact authorities you need to consult, followed by instructions for logging them in your first-flock operating plan.

Legal verification, property limits, and digging

Your first tier of resources governs what you are legally and physically allowed to do on your property. Always retrieve the primary written source; never rely on word of mouth.

Municipal clerk or city code portal: Start with your city or county clerk's office or the official municipal code database for your jurisdiction. This is where you locate the exact, currently enacted language that dictates how many hens you can keep, whether roosters are strictly banned, and what permits are required to keep livestock or poultry.

Zoning, planning, or building department: While the municipal code dictates whether hens are allowed, the zoning or building department usually dictates the physical structures. They hold the rules regarding coop setbacks from property lines, maximum roof heights, accessory-structure permits, and specific neighborhood zoning overlays that might supersede general city rules.

HOA, property management, or lease documents: Even if local municipal law allows chickens, your homeowner's association rules, deed restrictions, or lease agreements may still strictly prohibit them. You must consult your specific, current, legally binding documents. HOA and lease rules routinely override city permissions, and violating them can result in forced flock removal.

Animal control or local ordinance enforcement: Your local animal-control division, which typically operates under the police department or county sheriff, enforces noise limits, odor complaints, and loose-animal ordinances. They are also your primary authority to contact if a neighbor's loose dog threatens your flock.

811 or local utility-locate service Any digging guidance requires utility-locate verification. In the United States, contact 811 by phone or use your state 811 center's online system several business days before any digging, trenching, or ground-penetrating work, including small backyard projects. Wait for the required utility responses and markings, and check your state/local rules because timing, enforcement, and private-line locating responsibilities vary.

Animal health, biosecurity, and pest management

For the physical well-being of your birds and the specific environmental challenges of your suburban region, bypass internet searches and go straight to trusted agricultural and medical experts.

University Cooperative Extension: Most regions in the United States are served by a Cooperative Extension office tied to a land-grant university. These offices provide free, scientifically backed advice tailored to your climate. They are your primary resource for Integrated Pest Management (IPM) practices, offering safe guidelines for managing rodents, flies, manure, compost, and stormwater, as well as ensuring compliance with pesticide and rodenticide labels.

Poultry veterinarian: Locate an avian or livestock veterinarian who accepts backyard chickens before you have an emergency. Finding a qualified poultry vet is difficult in suburban settings where most clinics treat only dogs and cats. Do not rely on forums for diagnosis, prescribing, dosing medications, or home remedies.

State diagnostic laboratory and state veterinarian: Your state's department of agriculture employs a state veterinarian who monitors regional animal health. If you observe suspected contagious disease, sudden unexplained death, or multiple sick birds at once, escalate the issue immediately to your state diagnostic lab or state veterinarian. Do not transport sick or dead birds to a clinic or lab unless explicitly instructed to do so by a veterinarian, diagnostic lab, Extension office, state veterinarian, or public authority.

USDA APHIS (Animal and Plant Health Inspection Service): The United States Department of Agriculture's APHIS division monitors national disease threats, including Avian Influenza. USDA APHIS manages the National Poultry Improvement Plan (NPIP) and provides current biosecurity guidance, including the Defend the Flock program, which offers practical checklists for preventing the introduction of infectious diseases into your backyard.

Public health, food safety, and equipment safety

Because backyard chickens and humans share the same environment, hygiene, food safety, and physical safety are critical boundaries to maintain.

Centers for Disease Control and Prevention (CDC): The CDC is your primary public-health resource for managing the overlap between your flock and your household. They set the national guidance for preventing Salmonella and Campylobacter infections. The CDC dictates safe practices for handwashing, using dedicated coop footwear, and protecting vulnerable populations, specifically young children, older adults, pregnant people, and immunocompromised individuals.

State or local food-safety agencies: Food-safety guidance requires one conservative household default for egg collection, washing, and refrigeration. If you plan to gift, share, or sell eggs, check your state's department of agriculture or local health department. The rules for washing, refrigerating, packaging, reusing egg cartons, and labeling requirements vary dramatically by location and are strictly regulated.

Manufacturer safety manuals: Your primary authority for the safety of technical equipment is the manufacturer. Outdoor poultry electrical setup should follow the manufacturer instructions for each device, along with applicable local code or qualified electrical guidance. This includes ensuring heat plates, heated waterers, and fans are rated for their intended indoor/outdoor location. Use GFCI protection where required or recommended by manufacturer instructions, local electrical code, or a qualified electrician; ensure weather protection for connections; practice strict cord safety to reduce fire and trip hazards; and maintain child and pet boundaries. Do not attempt electrical modifications based only on DIY wiring advice or internet tutorials; use a qualified/licensed electrician when wiring changes are needed.

Documenting your sources

Take thirty minutes to verify and record the phone numbers, websites, office locations, and exact ordinance numbers for your municipality, county, and state. By gathering this accountability stack before you purchase birds, you establish a calm, reliable action plan for any legal questions, disease outbreaks, or safety concerns that arise during your flock's lifetime.

References & Sources

Chapter/topic-specific sources

Utility locating before digging

- Federal Communications Commission. FCC Designates 811 as the Nationwide Number to Protect Pipelines, Utilities from Excavation Damage (https://docs.fcc.gov/public/attachments/DOC-257293A1.pdf). Accessed 2026-06-13.

Animal abandonment legality and welfare

- Animal Legal & Historical Center, Michigan State University College of Law. Legal Protections for Chickens (https://www.animallaw.info/intro/legal-protections-chickens). Accessed 2026-06-13.

Chicken welfare and predator risk

- Oklahoma State University Extension. Protecting Small Poultry Flocks from Predators (https://extension.okstate.edu/fact-sheets/protecting-small-poultry-flocks-from-predators). Accessed 2026-06-13.

Reportable poultry disease signs and reporting

- USDA Animal and Plant Health Inspection Service. Defend the Flock | Signs of Illness (https://www.aphis.usda.gov/livestock-poultry-disease/avian/defend-the-flock/signs-illness). Accessed 2026-06-13.

Avian influenza response and spread

- USDA Animal and Plant Health Inspection Service. Avian Influenza (https://www.aphis.usda.gov/livestock-poultry-disease/avian/avian-influenza). Accessed 2026-06-13.

Virulent Newcastle disease response and spread

- USDA Animal and Plant Health Inspection Service. Disease Alert: Virulent Newcastle Disease (https://www.aphis.usda.gov/livestock-poultry-disease/avian/virulent-newcastle). Accessed 2026-06-13.

Reportable animal disease contacts

- USDA Animal and Plant Health Inspection Service. National List of Reportable Animal Diseases (https://direct.aphis.usda.gov/livestock-poultry-disease/surveillance/reportable-diseases). Accessed 2026-06-13.

Reportable poultry diseases

- USDA Animal and Plant Health Inspection Service. Voluntary 2026 U.S. National Animal Health Reporting System (NAHRS) Reportable Diseases, Infections, and Infestations List (https://direct.aphis.usda.gov/sites/default/files/nlrad-nahrs-disease-list.pdf). Accessed 2026-06-13.

Backyard poultry HPAI reporting

- University of Minnesota Extension. Avian influenza basics for urban and backyard poultry owners (https://extension.umn.edu/poultry-health/avian-influenza-basics-noncommercial-poultry-flock-owners). Accessed 2026-06-13.

Backyard poultry HPAI movement restrictions

- University of Maryland Extension. Backyard Poultry: HPAI Protection Guide (https://extension.umd.edu/sites/extension.umd.edu/files/2025-01/Backyard%20Poultry%20HPAI%20Protection%20Guide.S.Che%201.23.2025.pdf). Accessed 2026-06-13.

Backyard flock bird flu precautions

- Centers for Disease Control and Prevention. Backyard Flock Owners and Bird Flu (https://www.cdc.gov/bird-flu/risk-factors/backyard-flock-owners.html). Accessed 2026-06-13.

Private utility-line caveats

- Town of Leesburg, Virginia. Miss Utility - Call 811 Before You Dig (https://www.leesburgva.gov/departments/utilities-water-and-sewer/programs/call-811-before-you-dig-miss-utility). Accessed 2026-06-13.

Poultry illness reporting

- U.S. Department of Agriculture, Animal and Plant Health Inspection Service. Poultry Owners: Report Sick Birds! (https://www.aphis.usda.gov/sites/default/files/pos-hpai-report-sick-birds.508.pdf). Accessed 2026-06-13.

Backyard poultry disease and movement

- Merck Veterinary Manual. Common Infectious Diseases in Backyard Poultry (https://www.merckvetmanual.com/exotic-and-laboratory-animals/backyard-poultry/common-infectious-diseases-in-backyard-poultry?ruleredirectid=417). Accessed 2026-06-13.

Avian influenza and veterinary care

- Merck Veterinary Manual. Avian Influenza (Bird Flu) - Bird Owners (https://www.merckvetmanual.com/bird-owners/disorders-and-diseases-of-birds/avian-influenza-bird-flu). Accessed 2026-06-13.

Backyard poultry biosecurity

- Utah State University Extension. Backyard Poultry Biosecurity (https://extension.usu.edu/cyberlivestock/ou-files/BackyardPoultryBiosecurity.pdf). Accessed 2026-06-13.

Quarantine and poultry movement

- California Department of Food and Agriculture. Virulent Newcastle Disease Q&A for Veterinary Practitioners and Shelter Veterinarians (https://www.cdfa.ca.gov/ahfss/Animal_Health/pdfs/VND-QA-for-Veterinarians.pdf). Accessed 2026-06-13.

Backyard chicken veterinary care

- Texas A&M University School of Veterinary Medicine & Biomedical Sciences. Backyard Chickens Part 2: Maintaining Healthy Birds (https://vetmed.tamu.edu/news/pet-talk/backyard-chickens-part-2/). Accessed 2026-06-13.

Utility locating and 811

- Washington Utilities and Transportation Commission. Call Before You Dig (https://www.utc.wa.gov/public-safety/pipeline-safety/call-you-dig). Accessed 2026-06-13.

Rodenticide bait-station requirements

- U.S. Environmental Protection Agency. Restrictions on Rodenticide Products (https://www.epa.gov/rodenticides/restrictions-rodenticide-products). Accessed 2026-06-13.

Rodenticide bait-station safety

- U.S. Environmental Protection Agency. Choosing a Bait Station Product for Household Use (https://www.epa.gov/rodenticides/choosing-bait-station-product-household-use). Accessed 2026-06-13.

Rodenticide label and exposure safety

- U.S. Environmental Protection Agency. Safely Use Rodent Bait Products (https://www.epa.gov/rodenticides/safely-use-rodent-bait-products). Accessed 2026-06-13.

Tamper-resistant terminology

- U.S. Environmental Protection Agency. PRN 94-7: Label Improvement Program for the Revision of Use Directions for Commensal Rodenticides and Statement of the Agency's Policies on the Use of Rodenticide Bait Stations (https://19january2017snapshot.epa.gov/pesticide-registration/prn-94-7-label-improvement-program-revision-use-directions-commensal_.html). Accessed 2026-06-13.

Rodent control around poultry

- University of Maryland Extension. Rodent Control on Small Poultry Farms (https://extension.umd.edu/resource/rodent-control-small-poultry-farms-1). Accessed 2026-06-13.

Fly IPM around poultry

- Texas A&M AgriLife Extension. Poultry Pest Management - AgriLife Extension Entomology (https://extensionentomology.tamu.edu/resources/management-guides/poultry-pest-management/). Accessed 2026-06-13.

Trap placement and animal safety

- Washington State University Extension, Clark County. Natural Insect Control (https://wpcdn.web.wsu.edu/wp-extension/uploads/sites/2079/2025/01/4.1.31.25.natural-insect-controlAcrobat-Edit.pdf). Accessed 2026-06-13.

Insect trap safety and placement

- University of Florida IFAS Extension. ENY-2030/IN1103: Do-It-Yourself Insect Pest Traps (https://ask.ifas.ufl.edu/publication/IN1103). Accessed 2026-06-13.

Brooder heat-lamp safety

- University of Maine Cooperative Extension. Bulletin #2221, Simple Steps to Starting Healthy Chicks in Maine (https://extension.umaine.edu/publications/2221e/). Accessed 2026-06-13.

Brooder heat-lamp setup

- University of New Hampshire Extension. Brooding and Caring for Chicks (https://extension.unh.edu/resource/brooding-and-caring-chicks). Accessed 2026-06-13.

Heat-lamp fire hazard

- Central Whidbey Island Fire & Rescue. Chicken Coop Brood Lamp Fire Hazards (https://cwfire.org/chicken-coop-brood-lamp-fire-hazards/). Accessed 2026-06-13.

Backyard poultry zoonotic disease and biosecurity

- Centers for Disease Control and Prevention (CDC). Backyard Poultry (https://www.cdc.gov/healthy-pets/about/backyard-poultry.html). Accessed 2026-06-13.

Foodborne illness risk groups

- Centers for Disease Control and Prevention (CDC). People at Increased Risk for Food Poisoning (https://www.cdc.gov/food-safety/risk-factors/index.html). Accessed 2026-06-13.

Salmonella severity and complications

- Centers for Disease Control and Prevention (CDC). Clinical Overview of Salmonellosis (https://www.cdc.gov/sa

lmonella/hcp/clinical-overview/index.html). Accessed 2026-06-13.

Backyard poultry public-health precautions

- Wisconsin Department of Health Services. Backyard Poultry (https://www.dhs.wisconsin.gov/disease/backyard-poultry.htm). Accessed 2026-06-13.

Poultry biosecurity and quarantine

- U.S. Department of Agriculture, Animal and Plant Health Inspection Service (USDA APHIS). Biosecurity Education Kit (https://www.aphis.usda.gov/pest-disease/defend-flock/biosecurity-education-kit.pdf). Accessed 2026-06-13.

Marek's disease vaccination

- University of Maryland Extension. Recognizing and Preventing Marek's Disease in Small Flocks (https://extension.umd.edu/sites/extension.umd.edu/files/publications/FS-1007%20Recognizing%20and%20Preventing%20Mareks%20Disease%20in%20Small%20Flocks.pdf). Accessed 2026-06-13.

Backyard poultry infectious disease

- Merck Veterinary Manual. Common Infectious Diseases in Backyard Poultry (https://www.merckvetmanual.com/exotic-and-laboratory-animals/backyard-poultry/common-infectious-diseases-in-backyard-poultry). Accessed 2026-06-13.

Small flock care and biosecurity

- Colorado State University Extension, Boulder County Extension. Backyard and Small Flock Poultry (https://boulder.extension.colostate.edu/wp-content/uploads/sites/7/2020/04/Poultry-brochure-2020.pdf). Accessed 2026-06-13.

Baby chick pasty-butt care

- Purina Animal Nutrition. Pasty Butt: Sick Baby Chick Symptoms and Care (https://www.purinamills.com/chicken-feed/education/detail/pasty-butt-sick-baby-chick-symptoms-and-care). Accessed 2026-06-13.

Backyard chicken yard hazards

- N.C. Cooperative Extension, Richmond County Center. Extension@YourService: Is Your Backyard Chicken Proof? (https://richmond.ces.ncsu.edu/news/extensionyourservice-is-your-backyard-chicken-proof/). Accessed 2026-06-13.

Pesticide use around poultry

- University of Arizona Cooperative Extension. Backyard Chickens and Ectoparasites: Introduction and Management (https://extension.arizona.edu/sites/extension.arizona.edu/files/pubs/az1878-2021.pdf). Accessed 2026-06-13.

Excavation and underground installations

- Occupational Safety and Health Administration. 1926.651 - Specific Excavation Requirements. (https://www.osha.gov/laws-regs/regulations/standardnumber/1926/1926.651). Accessed 2026-06-13.

Chicken coop electrical and fire safety

- National Fire Protection Association. Backyard Chicken Coop Safety (https://www.nfpa.org/-/media/project/storefront/catalog/files/safety-tip-sheets/chickencoopsafetytips.pdf?rev=7b826dd3a9004d3ebe83c70a053904d8). Accessed 2026-06-13.

Chick brooder heat source safety

- University of Maine Cooperative Extension. Simple Steps to Starting Healthy Chicks in Maine (https://extension.umaine.edu/publications/wp-content/uploads/sites/52/2020/04/2221-printable.pdf). Accessed 2026-06-13.

Home electrical fire risk

- U.S. Fire Administration. Appliance and Electrical Fire Safety (https://www.usfa.fema.gov/prevention/home-fires/prevent-fires/appliance-and-electrical/). Accessed 2026-06-13.

Electrical safety and qualified electricians

- Electrical Safety Foundation International. Home Safety (https://www.esfi.org/home-safety). Accessed 2026-06-13.

Egg washing and shell protection

- U.S. Department of Agriculture, Agricultural Marketing Service. Egg-Grading Manual (https://www.ams.usda.gov/sites/default/files/EggGradingManual.pdf). Accessed 2026-06-13.

Backyard flock egg handling

- Penn State Extension. How to Safely Handle and Store Eggs from Your Home Flock (https://extension.psu.edu/how-to-safely-handle-and-store-eggs-from-your-home-flock). Accessed 2026-06-13.

Salmonella reduction in egg handling

- University of Minnesota Extension. Reducing salmonella risk in table egg production (https://extension.umn.edu/small-scale-poultry/reducing-salmonella-risk-table-egg-production). Accessed 2026-06-13.

Consumer egg safety

- U.S. Food and Drug Administration. What You Need to Know About Egg Safety (https://www.fda.gov/food/buy-store-serve-safe-food/what-you-need-know-about-egg-safety). Accessed 2026-06-13.

Bird equipment cleaning and cross-contamination

- Centers for Disease Control and Prevention, Healthy Pets, Healthy People. Birds (https://www.cdc.gov/healthy-p

ets/about/birds.html). Accessed 2026-06-13.

Compost treatment standards

- U.S. Food and Drug Administration. Frequently Asked Questions on FSMA (https://www.fda.gov/food/food-safety-modernization-act-fsma/frequently-asked-questions-fsma). Accessed 2026-06-13.

Produce safety and manure compost

- Cornell CALS, National Good Agricultural Practices Program. Soil Amendments (https://cals.cornell.edu/national-good-agricultural-practices-program/resources/educational-materials/decision-trees/soil-amendments). Accessed 2026-06-13.

Backyard composting

- U.S. Environmental Protection Agency. Composting At Home (https://www.epa.gov/recycle/composting-home). Accessed 2026-06-13.

Poultry manure pathogen reduction

- Penn State Extension. Composting Poultry Manure to Reduce the Spread of Pathogens (https://extension.psu.edu/composting-poultry-manure-to-reduce-the-spread-of-pathogens/). Accessed 2026-06-13.

Backyard flock egg washing

- University of Georgia Alumni Association / UGA Cooperative Extension. Backyard Flock Tips on National Egg Day (https://alumni.uga.edu/2019/06/03/backyard-flock-tips-on-national-egg-day/). Accessed 2026-06-13.

Poultry mycotoxins and moldy feed

- Merck Manual Veterinary Manual. Mycotoxicoses in Poultry - Poultry - Merck Veterinary Manual (https://www.merckvetmanual.com/poultry/mycotoxicoses-in-poultry/mycotoxicoses-in-poultry). Accessed 2026-06-13.

Mycotoxins in animal food

- U.S. Food and Drug Administration, Center for Veterinary Medicine. Chemical Contaminants (https://www.fda.gov/animal-veterinary/biological-chemical-and-physical-contaminants-animal-food/chemical-contaminants). Accessed 2026-06-13.

Small flock disease prevention

- University of Maryland Extension. Disease Management (https://extension.umd.edu/resource/disease-management). Accessed 2026-06-13.

Poultry botulism

- Merck Manual Veterinary Manual. Botulism in Poultry - Poultry - Merck Veterinary Manual (https://www.merckvetmanual.com/poultry/botulism/botulism-in-poultry). Accessed 2026-06-13.

Backyard poultry botulism prevention

- Small and Backyard Poultry, Cooperative Extension. BOTULISM IN POULTRY (https://poultry.extension.org/articles/poultry-health/common-poultry-diseases/botulism-in-poultry/). Accessed 2026-06-13.

Avian botulism in birds

- U.S. Geological Survey. Avian Botulism (https://www.usgs.gov/diseases-of-terrestrial-wildlife/avian-botulism). Accessed 2026-06-13.

Animal handling and human safety

- Centers for Disease Control and Prevention. Hygiene Practices Around Animals | Water, Sanitation, and Environmentally Related Hygiene (WASH) | CDC (https://www.cdc.gov/hygiene/about/hygiene-practices-around-animals.html). Accessed 2026-06-13.

Handwashing technique

- Centers for Disease Control and Prevention. About Handwashing | Clean Hands | CDC (https://www.cdc.gov/clean-hands/about/index.html). Accessed 2026-06-13.

Safety, legal, regulatory, or technical sources

- U.S. Department of Transportation. Call 811 Before You Dig (https://www.transportation.gov/content/call-811-you-dig-1). Accessed 2026-06-13.
- USDA Animal and Plant Health Inspection Service. Avian Influenza (https://www.aphis.usda.gov/livestock-poultry-disease/avian/avian-influenza). Accessed 2026-06-13.
- U.S. Department of Agriculture, Animal and Plant Health Inspection Service. Poultry Owners: Report Sick Birds! (https://www.aphis.usda.gov/sites/default/files/pos-hpai-report-sick-birds.508.pdf). Accessed 2026-06-13.
- Merck Veterinary Manual. Common Infectious Diseases in Backyard Poultry (https://www.merckvetmanual.com/exotic-and-laboratory-animals/backyard-poultry/common-infectious-diseases-in-backyard-poultry?ruleredirectid=417). Accessed 2026-06-13.
- California Department of Food and Agriculture. Virulent Newcastle Disease Q&A for Veterinary Practitioners and Shelter Veterinarians (https://www.cdfa.ca.gov/ahfss/Animal_Health/pdfs/VND-QA-for-Veterinarians.pdf). Accessed 2026-06-13.
- Minnesota Department of Public Safety. Private utilities (https://dps.mn.gov/divisions/ops/homeowners/private-utilities). Accessed 2026-06-13.
- U.S. Environmental Protection Agency. PRN 94-7: Label Improvement Program for the Revision of Use Directions for Commensal Rodenticides and Statement of the Agency's Policies on the Use of Rodenticide Bait Stations (https://19january2017snapshot.epa.gov/pesticide-registration/prn-94-7-label-improvement-program-revision-use-directions-commensal_.html). Accessed 2026-06-13.
- Central Whidbey Island Fire & Rescue. Chicken Coop Brood Lamp Fire Hazards (https://cwfire.org/chicken-coop-brood-lamp-fire-hazards/). Accessed 2026-06-13.

- Centers for Disease Control and Prevention (CDC). Backyard Poultry (https://www.cdc.gov/healthy-pets/about/backyard-poultry.html). Accessed 2026-06-13.
- Wisconsin Department of Health Services. Backyard Poultry (https://www.dhs.wisconsin.gov/disease/backyard-poultry.htm). Accessed 2026-06-13.
- Merck Veterinary Manual. Common Infectious Diseases in Backyard Poultry (https://www.merckvetmanual.com/exotic-and-laboratory-animals/backyard-poultry/common-infectious-diseases-in-backyard-poultry). Accessed 2026-06-13.
- Occupational Safety and Health Administration. 1926.651 - Specific Excavation Requirements. (https://www.osha.gov/laws-regs/regulations/standardnumber/1926/1926.651). Accessed 2026-06-13.
- National Fire Protection Association. Backyard Chicken Coop Safety (https://www.nfpa.org/-/media/project/storefront/catalog/files/safety-tip-sheets/chickencoopsafetytips.pdf?rev=7b826dd3a9004d3ebe83c70a053904d8). Accessed 2026-06-13.
- U.S. Fire Administration. Appliance and Electrical Fire Safety (https://www.usfa.fema.gov/prevention/home-fires/prevent-fires/appliance-and-electrical/). Accessed 2026-06-13.
- U.S. Department of Agriculture, Agricultural Marketing Service. Egg-Grading Manual (https://www.ams.usda.gov/sites/default/files/EggGradingManual.pdf). Accessed 2026-06-13.
- U.S. Food and Drug Administration. What You Need to Know About Egg Safety (https://www.fda.gov/food/buy-store-serve-safe-food/what-you-need-know-about-egg-safety). Accessed 2026-06-13.
- Centers for Disease Control and Prevention, Healthy Pets, Healthy People. Birds (https://www.cdc.gov/healthy-pets/about/birds.html). Accessed 2026-06-13.
- U.S. Environmental Protection Agency. Composting At Home (https://www.epa.gov/recycle/composting-home). Accessed 2026-06-13.
- Merck Manual Veterinary Manual. Mycotoxicoses in Poultry - Poultry - Merck Veterinary Manual (https://www.merckvetmanual.com/poultry/mycotoxicoses-in-poultry/mycotoxicoses-in-poultry). Accessed 2026-06-13.
- U.S. Food and Drug Administration, Center for Veterinary Medicine. Chemical Contaminants (https://www.fda.gov/animal-veterinary/biological-chemical-and-physical-contaminants-animal food/chemical-contaminants). Accessed 2026-06-13.
- U.S. Geological Survey. Avian Botulism (https://www.usgs.gov/diseases-of-terrestrial-wildlife/avian-botulism). Accessed 2026-06-13.
- Centers for Disease Control and Prevention. About Handwashing | Clean Hands | CDC (https://www.cdc.gov/clean-hands/about/index.html). Accessed 2026-06-13.

Further reading and local-resource note

Rules, recommendations, and professional guidance can change and may vary by jurisdiction. Confirm current local requirements and seek qualified expert advice where the book identifies that need.

www.ingramcontent.com/pod-product-compliance
Lightning Source LLC
LaVergne TN
LVHW062029051126
851858LV00006B/125
* 9 7 8 1 9 6 3 1 4 2 5 6 3 *